Quincy Adams Gillmore

Notes on the Compressive Resistance of Freestone,

brick piers, hydraulic cements, mortars and concretes

Quincy Adams Gillmore

Notes on the Compressive Resistance of Freestone,
brick piers, hydraulic cements, mortars and concretes

ISBN/EAN: 9783337286026

Printed in Europe, USA, Canada, Australia, Japan

Cover: Foto ©berggeist007 / pixelio.de

More available books at **www.hansebooks.com**

NOTES

ON THE

COMPRESSIVE RESISTANCE

OF

FREESTONE, BRICK PIERS, HYDRAULIC CEMENTS, MORTARS AND CONCRETES.

BY

Q. A. GILLMORE, Ph.D.,

*Colonel Corps of Engineers, Brevet Major-General, U. S. A.; Author of
"Treatise on Limes, Cements, etc.;" "Treatise on Coignet Béton and
Artificial Stones;" "Report on Compressive Strength, etc.,
of the Building Stones in the U. S.;" "Treatise
on Roads, Streets and Pavements,"
etc. etc. etc.*

NEW YORK:

JOHN WILEY & SONS,

15 ASTOR PLACE.

1888.

PREFATORY NOTE.

THE tests of the several kinds of building materials dis-
ussed in the following pages were obtained mostly by a ma-
hine of extreme delicacy, having a maximum working pressure
.f 800,000 pounds. It was erected at the Watertown Arsenal,
ıear Boston, some years ago, by Mr. Albert H. Emery, under
he direction of the Board on Iron and Steel appointed by the
'resident in accordance with the Act of Congress of March 3,
875.

I desire to acknowledge my obligations to Lieut.-Colonel
?. H. Parker, Ordnance Department U. S. Army, command-
ng Watertown Arsenal, for the active interest taken by him in
he tests, and his ever-ready assistance in promoting the work ;
ınd to Mr. J. E. Howard, the engineer of the testing-machine,
vhose acknowledged ability in operating the ponderous instru-
nent was most skilfully applied in carrying the experiments
o a successful conclusion.

My principal professional indebtedness is due to Mr. John
_. Suess. Senior Assistant Engineer in my office. To him
s due in very large measure that untiring energy, unflag-
;ing patience, and scientific discussion of the various problems
nvolved which so greatly contributed to final success.

It is not too much to assert, that without his zealous coöp-
:ration the work would have been suspended at a stage far
;hort of completion.

Q. A. G.

CONTENTS.

COMPRESSIVE RESISTANCE

OF

FREESTONE, BRICK PIERS, HYDRAULIC CEMENTS, ETC.

CHAPTER I.

INTRODUCTION.

CERTAIN tests for ascertaining the compressive strength of building material were carried on under my direction about twelve years ago, and a preliminary report, dated August 10, 1875, was printed as Appendix II. of the Annual Report of the Chief of Engineers for 1875. A new series of experiments was made toward the close of the year 1883, for the purpose of obtaining further information in regard to the resistance and behavior under compressive strains, of hydraulic cement, of mortars and concretes made with cement, of brick piers, and of freestone, either in the form of cubes of various sizes, or of prisms square in cross-section, but of less height than corresponding cubes.

The earlier tests were made with a hydraulic press whose indicated pressure did not exceed 100,000 pounds. The dimensions of the specimens that were tested were therefore necessarily restricted. A few 11-inch cubes of Berea sandstone were crushed by means of a 2000-ton press at the Brooklyn Navy Yard, but the results were not thought to have much weight, as the accuracy of the testing-machine was doubted.

The chief object of these earlier investigations was to determine the compressive strength, specific gravity, and ratio of absorption of the most commonly used building-stones of the United States. The average results obtained from specimens of 216 different kinds of granite, marble, limestone, and sandstone were given in a general table appended to my report of August 10, 1875. The specimens were 2-inch cubes, and were crushed between cushions or disks of soft pine-wood three eighths of an inch thick. One of these cushions was placed under the bottom face of the cube, the other on top.

A number of special tests were also reported.

They were made to determine the effects of changing the nature of the pressing-surfaces between which the specimens were tested; of varying the relation between the heights of specimens and the areas of their bed-faces; and of changing the absolute dimensions of cubes of the same material.

It was found that when steel or wood formed the pressing-surfaces the phenomena of breakage were nearly the same. Generally there were two characteristic fragments more or less pyramidal in form, with a portion of the bed-faces as bases, and with lateral angles of about 45 degrees; with steel plates there sometimes appeared to be a tendency to form but one pyramid, with lateral angles of approximately 60 degrees. With wood, the end-pieces seemed to be slightly more prismoidal; with steel, more wedge-shaped. The final destruction of a specimen was generally accompanied by a loud report.

Different results were obtained when lead or lace-leather was interposed between the specimens and the pressing surfaces. At the moment of fracture numerous cracks, parallel to the direction of pressure and perpendicular to the compressed bed-faces, appeared upon the sides of the specimen, and its cohesion was destroyed almost instantaneously. The fragments were prismatic, their greatest dimension or length being parallel to the direction of the pressure. A comparatively large amount of stone-dust was produced at the same time. The action of the lead cushions was ascribed to the capacity of that metal to flow when under sufficient pressure. The side of the lead cushion next to the steel plate of the testing-machine is made smooth, the other side is driven by the pressure into the minute interstices and depressions of the stone, forming innumerable wedges which tend to split it, while the normal pressure acts powerfully to open it in the middle. At the moment of fracture a faint dull report could generally be heard; occasionally no audible sign was given announcing the destruction of the sample.

Three different series of tests were made to ascertain the effect of applying cushions of various materials. In the first two series, all stones crushed were in the form of 2-inch cubes; in the third series, one set consisted of 1½-inch cubes, the other of 2-inch cubes.

The results obtained may be briefly recapitulated as follows:

First Series.—With notably tough and first-class building-stones, such as Millstone Point granite, East Chester marble, and blue Berea sandstone, the average crushing resistances were found to be in the following proportion, the leather having been tried with sandstone only: steel, 100; wood, 94; lead, 65; leather, 60.

Second Series.—The second series of tests was made upon stones having nearly or quite as compact and close a texture on the ground surface as those of the first series, but which were more friable upon the surface of fracture, and evidently possessed less cohesive and tensile strength. These were samples of Keene granite, and of a Vermont marble—a clear, smooth, delicate-looking stone. The following ratios were obtained: steel, 100; wood, 82; lead, 65; leather, 63.5.

Third Series.—The third series of tests was made with stone which was so soft that wood did not sensibly spread, nor lead or leather flow under such comparatively low pressures as were sufficient to crush the specimens; in other words, it was expected that steel, wood, lead, and leather would, at some low point of crushing pressure, give approximately identical results.

For this purpose, Sebastopol limestone (a species of chalk), a soft kind of sandstone, and two sets of cubes of Massillon sandstone were tried.

The following ratios were obtained with them:

| Kind of Stone. | Ratio of Resistance with Cushions of | | | | Remarks. |
	Steel.	Wood.	Lead.	Leather.	
Sebastopol limestone.....	100	100	100	100	Mean of fourteen 2-inch cubes.
Drab-colored sandstone..	100	100	100	...	Mean of three 1½-inch cubes.
Massillon sandstone......	100	110	90	59.4	Mean of sixteen 2-inch cubes.
" "	100	103	85	...	Mean of five 2-inch cubes.

This table shows about equal crushing resistance with steel and wood, but the actual compressive strength of the stones of the third series was much below that of the granite and marble of the first and second series.

From these experiments it was inferred that with stones combining considerable hardness with toughness, steel and wood give approximately equal results; that with stones which, though hard, are yet deficient in toughness, the peculiar action of wood cushions, which spread sideways and thus produce strains requiring tensile resistance, causes the stone to be crushed under a smaller load than with steel, which tends to bind the stone together by its rigidity and frictional resistance to lateral yielding; and that in decidedly soft stones the ability of a specimen to resist crushing is overcome before sufficient pressure is developed to spread the wood fibres, or to make the lead flow.

The relative resisting power of stone prisms, square in cross-section, but of various heights, was investigated at about the same time, blue Berea sandstone being used for this purpose.

Broken between steel plates, the ultimate strength of a 1-inch cube averaged 9500 pounds; four isolated cubes of the same size and kind would therefore have yielded under an aggregate load of 38,000 pounds. The same amount of material formed as a solid slab, 2 inches square and 1 inch high, developed an average crushing resistance of nearly 76,000 pounds (more precisely, 75,888 pounds), or twice as much as the set of four 1-inch cubes having an aggregate bed-area exactly equal to that of the single slab.

Two-inch cubes broke under an average load of nearly 50,000 pounds. Samples with the same bed-area or cross-section, but with twice the height of a cube, sustained a mean pressure of not quite 44,000 pounds.

Similar results were obtained with specimens $1\frac{1}{2}$ inches square in cross-section. When $\frac{3}{4}$ of an inch high, the samples were crushed under an average load of 34,643 pounds; in the form of cubes, under a load of 25,350 pounds; and when 4 inches high, under a load of 22,432 pounds.

When similar samples were broken between wooden cushions, the difference of strength in favor of slabs was much less marked than when the crushing was done between steel plates, for reasons already suggested.

The results of the tests seemed to indicate not only that slabs increase in resistance, per square inch, as their surfaces increase, but also that the strength per square inch of cross-section of cubes increases with their size, although in a lesser ratio. To investigate this latter question, a series of experiments was made upon various-sized cubes composed of two kinds of Berea stone. In one set, made of a yellowish-gray stone, the sides of the cubes increased from one quarter of an inch to four inches; in the other set, of bluestone, the sides of the cubes varied from one inch to two inches and three quarters. The sides of the cubes increased successively by quarter inches. The first set was broken between wooden cushions; the second set, a harder variety of stone, between steel plates.

A curve was constructed for each set, the sides of the cubes in inches being the abscissas, and the crushing load of each specimen, in pounds per square inch of bed-surface, the ordinates. In other words, the ordinate for any specimen was the quotient of the total compressive resistance of the cube divided by the number of square inches in one of its faces. It was found that the approximate form of the theoretical curve was that of a cubic parabola, with the equation

$$y = a\sqrt[3]{x},$$

in which a is the pressure in pounds required to crush a 1-inch cube, x the side of any cube expressed in inches, and y the pressure in pounds per square inch of bed-surface needed to crush it.

These experiments seemed to indicate that *with cubes of the same material the crushing resistance per square inch of compressed surface increases, approximately, in the ratio of the cube roots of the sides of the respective cubes.*

Since it was unsafe to work the press then used beyond 100,000 pounds, the size of the specimens of the harder or blue Berea stone was restricted to 2¾-inch cubes; of the softer kind to 4-inch cubes. The range of the experiments was therefore too limited to justify the assumption that the formula deduced from them would prove sufficiently correct when applied to larger cubes. It was noted at the time that the formula was not borne out by the results obtained with five 11-inch cubes of Berea stone that were crushed at the Brooklyn Navy Yard. They gave way at somewhat less recorded pressure per square inch of bed-surface than 2-inch cubes of the same stone.

The question whether there is a gradual increase or decrease of compressive strength per square inch of pressed surface, as the size of cubes of the same kind and quality of stone or similar building material increases, was therefore still unsettled, and had to remain so until a more powerful testing-machine became available.

CHAPTER II.

OBJECT OF EXPERIMENTS. AND CHARACTER AND FORM OF SPECIMENS TESTED.

In 1875, the President of the United States, under an Act of Congress approved March 3, 1875, appointed a Board composed of Army and Navy officers and civil engineers, who were authorized to secure a testing-machine with which to make tests of "iron, steel, and other metals." This board in the same year entered into a contract with Mr. A. H. Emery to construct and erect at the Watertown Arsenal, near Boston, Mass., a 400-ton testing-machine, to be used for determining the tensile and compressive strength of material entering into engineering and architectural structures.

The machine was completed in February, 1879, and soon became known as the most perfect and reliable machine of its kind in existence, as it combined great power with extraordinary delicacy of weighing apparatus.

It was decided to extend the former experiments with this new and more powerful machine.

It was thought best to select materials possessing as uniform texture as practicable, in order to exclude, if possible, disturbing influences resulting from the different nature, size, and unequal distribution of individual grains.

In addition to uniformity of texture or grain, the degree of hardness and toughness was considered. The cubes of each kind of material were to increase, by certain increments, from one, two, or four inches on a side, as the case might be, to as large a size as would presumably resist nearly the entire power of the machine. It was obviously desirable to vary the sizes of the cubes between as wide limits as possible.

It was therefore unwise to employ cubes of the harder classes of natural building stone, such as granite, syenite, etc., as the capacity of the machine would be exceeded by cubes of comparatively small size.

Former examinations and tests of the softer varieties of building material suggested a variety of red sandstone known as Haverstraw freestone. This kind of stone, in the form of 2-inch cubes, had been found to yield under an average load of 4350 pounds per square inch of bed-surface, and the grain, though somewhat coarse, appeared to be rather uniform.

Cubes of this material varying, by increments of an inch, from one inch to twelve inches on a side were prepared, four cubes of each size being made. Two sets of prisms, square in cross-section and with varying heights less than that of corresponding cubes, were also prepared. One set measured $4'' \times 4''$ on the bed-surface, the other $8'' \times 8''$. Each sample of sandstone was wrought to its proper form by a skilled stone-cutter and the bed-faces were rubbed plane.

Cubes and prisms of neat cement were prepared, in order that a material presumably of as nearly homogeneous texture as practicable might be tested. A quantity of Dyckerhoff's Portland cement (from Amoeneburg on the Rhine, Germany) being on hand, this brand was employed. The sides of the cubes made of this cement varied by increments of an inch from one inch to twelve inches. There were six samples of each size. To these were added three sets of square prisms of less height than corresponding cubes ; their bed-faces measuring $4'' \times 4''$, $8'' \times 8''$ and $12'' \times 12''$ respectively.

As little water as practicable was used in preparing the cement for the moulds. The moulds were boxes of pine wood, without top or bottom, smooth inside and held together by bolts passing through opposite sides beyond the ends. The bottom was formed by placing the mould upon a smooth bluestone flag, and the interior of the box was well greased to prevent adhesion of the damp material. The moistened cement was put into the box and gradually consolidated by tamping, using a hammer of about four pounds weight, and a follower consisting of a short stick of hard wood.

The blocks were taken from the moulds as soon as they could be safely handled, the smallest a short time after being formed, the largest in about twelve hours. They were then buried in sand on the floor of one of the casemates of Fort Tompkins, not only to keep them moist, but as a precaution against frost and changes of temperature generally. They remained there until taken to the Watertown Arsenal to be tested.

A number of mortar and concrete cubes of various sizes were made, using different brands of American cements.

Of the brand known as Norton's cement, four different sets of cubes were made. Each set comprised duplicate cubes of the dimensions generally of 4 inches, 6 inches, 8 inches, 12 inches, and 16 inches on the edge. Their composition was as follows:

First Set.—Cubes of mortar: proportion, 1 vol. cement paste, 1½ vols. sand.

Second Set.—Cubes of concrete: proportion, 1 vol. cement paste, 1½ vols. sand, and 6 vols. broken stone.

Third Set.—Cubes of mortar: proportion, 1 vol. cement paste, 3 vols. sand.

Fourth Set.—Cubes of concrete: proportion, 1 vol. cement paste, 3 vols. sand, and 6 vols. broken stone.

Two sets of mortar and concrete cubes, corresponding as to sizes and numbers of blocks to those of Norton's cement, were made of the brand known as National Portland cement.

First Set.—Cubes of mortar: proportion, 1 vol. cement paste, and 3 vols. sand.

Second Set.—Cubes of concrete: proportion, 1 vol. cement paste, 3 vols. sand, and 6 vols. broken stone.

Two sets of mortar and concrete cubes were prepared with the cement known in market as the Newark Company's Rosendale cement.

The first set was formed of mortar, in the proportion of 1 vol. cement, dry measure, to 3 vols. sand. It comprised duplicate cubes, varying by increments of 2 inches from 2 inches to 16 inches on a side.

The second set was made of concrete, in the proportion of 1 vol. cement, dry measure, 3 vols. sand, 2 vols. gravel, and 4 vols. broken stone. It comprised duplicate cubes, varying by increments of 2 inches from 4 inches to 18 inches on a side.

In preparing the mortar, the cement paste was first made with as little water as practicable; to this the sand was added, thus forming a stiff mortar. For concrete blocks, gravel and broken stone were added in the requisite proportions, and the whole mass was thoroughly worked and mixed. In some instances when needed, the broken stone and gravel were first dampened by slightly sprinkling with water. The moulds were of the same kind as used for the cubes of neat cement. The material in the larger moulds was consolidated by ramming with a conical-pointed iron rammer of about eight pounds weight, two feet in length, and one inch in diameter. A lighter rammer was used for the smaller blocks.

Silicious, fresh-water sand was used in making the mortars. The broken stone for the concretes was of nut size, angular and sharp-edged, and consisted of a gray variety of hard and tough limestone.

All of the mortar and concrete blocks were kept buried in sand in a casemate of Fort Tompkins until they were shipped to the place of testing.

Incidentally it was thought desirable to make a few tests of the crushing strength of brick in the form of short piers. Six piers were built, each about 12 inches (1½ brick) square in cross-section, and six courses in height, with a strong bluestone flag at either end. Common hard North River bricks were used, averaging about 8 inches in length, 3½ inches in width, and 2¼ inches in thickness. The mortar was made of one part of the Newark Company's Rosendale cement and two parts of sand. No special care was taken in building the piers, as it was intended that they should represent ordinary, average brickwork. The mortar joints averaged about ⅜ths of an inch in thickness.

The blocks made of Dyckerhoff's and of the Newark Company's Rosendale cement were from 1 year 10 months to 1 year 11 months old when crushed; the brick piers had nearly the same age; the cubes made with Norton's and National Portland cement were about 3 years 10 months old. The exact age of each sample when broken is given in the accompanying general tables.

CHAPTER III.

DESCRIPTION OF TESTS.

In ascertaining the compressive strength of columns or prisms with flat, square ends, it is necessary that the two end-surfaces should be parallel to each other, and that these surfaces should be smooth and plane. It is extremely difficult, if not practically impossible, to dress and finish natural stone or to mould artificial stone so accurately as to fulfil strictly these conditions, and the difficulty increases with the size of the specimen.

The pressing-surfaces of the heads of both the stationary and the movable holder of the Watertown machine, one of them being of gun-iron, the other of cast steel, are as truly plane and smooth as the best mechanical skill can make them; they are finished to a degree which cannot be attained with relatively coarse-grained material such as freestone, cement, mortars, and concrete. The movable holder of the straining-press had a strong adjustable head-plate, by means of which the bed-surfaces of those test-pieces whose ends were not truly parallel could be brought into close contact with the faces of the holder-plates.

Another difficulty became manifest soon after beginning the testing operations. The cubes of neat cement which were first subjected to testing had been prepared with great care, but in a number of instances it was noticed that their beds were not in contact with the holder-plates at all points, in consequence of their being either slightly warped, rounded, or otherwise deficient. These irregularities were in reality very slight, and would not have been of any importance in practical work, but it was decided that they could not be ignored when comparing the strength of various-sized samples of the same material. Since similar irregularities were observed in a number of samples of freestone, and in the mortar and concrete blocks, some method of finishing off the upper and lower bed-faces, so as to secure plane and parallel surfaces, had to be devised.

A preliminary trial was made with a 3-inch cube of neat cement, one bed of which was somewhat deficient. It was put in a lathe and faced with a steel cutter. The result was satisfactory; but it became apparent that this method of treating many samples, especially the larger ones, would be objectionably slow, inasmuch as the cutter wore out very rapidly.

The use of an emery-wheel was then suggested, and experiments were made with one small sample of each kind of material. Satisfactory results were obtained with the cement blocks, but the surface was glazed; the freestone was tolerably well finished, but when tried on mortar and concrete the process failed.

The experiments having been partially successful, it seemed desirable to rig up a large lathe at the arsenal with the necessary machinery for mounting a 14-inch emery-wheel to face deficient cubes of freestone and cement measuring as much as 12 inches on a side, although the mortars and concretes would have to be treated differently. The plan had to be abandoned, however, as the lathe was otherwise employed, and could not be spared for this purpose.

The method previously followed at the Watertown Arsenal when testing the crushing strength of brick piers, under direction of Colonel T. T. S. Laidley, late commanding officer at the arsenal, was next tried. Those piers were hoisted into position between the pressure-heads of the testing-machine, which just touched their end-faces. The joints of the bottom and of the two vertical sides (the pier lying horizontally, as required by the construction of the testing-machine) were first closed with a stiff paste of plaster of Paris; when the plaster joints were dry and hard, semi-fluid plaster paste was poured in at the top joints until every cavity between the pier-head and iron plate was thought to be filled. The plaster was allowed to harden for 24 or 36 hours, and the pressure then put on.

This process would of course have been too tedious where many cubes and prisms had to be tested, but the advantage of finishing off the beds with a thin coating of plaster paste, which gave them a smooth surface corresponding to that of the pressing-plates of the machine, was obvious.

The addition of a plaster coating of such minute thickness could not, in any appreciable degree, modify the behavior of the specimen while being compressed.

The actual method adopted was as follows : Some large, heavy, smoothly-planed cast-iron plates were procured, and placed horizontally upon low supports resting upon the floor of one of the shops of the arsenal. The upper surface of each plate was oiled, and a thin layer of rather stiff paste of plaster of Paris poured upon it. The face of the cube or prism to be plastered was next washed with diluted paste ; the piece was then carefully placed upon the iron plate, pressing it firmly into the plaster bed. It remained there undisturbed for about half an hour, and was then lifted off ; a thin layer or skin of plaster adhered to the face of the piece, presenting a smooth, plane, and marble-like surface. The opposite face was then similarly treated. The length of the piece, from bed to bed, was carefully measured to the nearest one-hundredth of an inch, both with and without plaster. The dimensions of its cross-section were taken in like manner. The plaster was allowed to harden for about 36 or 48 hours before the sample was tested.

In the case of all of the mortar cubes and of half of the concrete cubes made with the Newark Company's Rosendale cement, cushions of pine-wood were interposed between the plastered heads of the specimen and the machine-heads. The use of such cushions was dispensed with while testing the other kinds of material.

While ascertaining the crushing strength of specimens, the rate of compression as the load was gradually increased was also measured in a number of cases.

The amount of compression or extension of the specimen was measured by a micrometer designed by Mr. J. E. Howard, the engineer of the testing-machine. This instrument consists essentially of two flat bars, holding between them a little arbor upon which a graduated circle or limb is mounted. One end of one bar is clamped to the movable holder of the straining-press, and the farther end of the other bar to the stationary holder of the machine. As soon as compression begins, the movable holder moves towards the stationary holder, carrying the bar which is clamped to it in the same direction; the arbor being held tightly between the two bars is made by friction to rotate, carrying with it the circular limb. The graduation reads to one-thousandth of an inch; but a practised eye can estimate ten-thousandths of an inch with considerable accuracy.

This micrometer was used in all tests of samples of eight inches in height and upwards.

Since the testing-machine is so constructed that the moving force, whether applied for tension or for compression, acts in a horizontal direction, some pressure must be applied for the purpose of holding the specimen in its proper position between the machine-heads. An initial pressure of 5000 pounds was put on for holding the larger cubes, and a less pressure for the smaller or weaker samples. At this initial pressure the graduated limb was set at zero.

As the load was gradually increased, the amount of compression was read off and noted. At certain intervals the strain was relaxed, returning to the initial pressure. The set, if any, was noted, and the straining-press again put to work.

The results of these micrometer measurements for compression and set are given in Special Tables I. to X., and in the diagram sheets I. to VIII. accompanying this report.

To facilitate comparison of the curves of compression they are all drawn to the same scale; the ordinates representing the pressure in pounds, and the abscissas the amount of compression in inches. With few exceptions the diagrams show that during the first stages of applying the pressure the compression of the piece takes place at a comparatively rapid and uneven rate. The curve is irregular, and more or less convex toward the axis of abscissas. As the load increases the curve gradually straightens, and later on becomes concave, inclining toward the horizontal axis. This concavity is much more marked with the mortars and concretes than with the cements and freestone. In discussing the results obtained with the several kinds of material tested, the phenomena attending compression and set will be briefly considered.

CHAPTER IV.

TESTS OF HAVERSTRAW FREESTONE.

THIS stone belongs to the class known as brownstone, its color being a warm and somewhat dark reddish-brown. It is of moderate fineness of grain, and apparently rather homogeneous in texture. In some instances, however, samples after fracture showed distinct traces of lamination, thin seams or strata of coarser grain parallel to the bed being visible. The average weight of this material was about 136.5 pounds per cubic foot, the specific gravity being 2.184.

PHENOMENA ATTENDING FAILURE OF SPECIMENS.

The usual manner in which cubes of amorphous stone fail under a crushing load was again illustrated by this material. The principal fragments generally consisted of two irregular pyramids, more or less fully developed, with the bed-faces, or rather the larger portion of the same, as bases. The lateral parts of the cubes were forced off the sides of the pyramidal core, forming occasionally comparatively large slabs. One or two of the sides of a cube sometimes split off nearly entire; but as a rule they broke off in smaller fragments. The material remaining between these fragments and the pyramids was well disintegrated, and partially ground to powder of various degrees of fineness. In several cases but one pyramid was fairly developed—apparently at the expense of the opposite one. In numerous instances the two pyramids remained loosely connected after fracture, having the appearance of sliding past each other, instead of abutting with their apexes. This condition was occasionally modified by one pyramid seeming to pierce the other, leaving in the latter, when the former was detached from it, a crater-like recess, as shown in sketch, the dotted areas in which represent the lateral pieces and ground material broken off at the moment of fracture. It seems as if the cube had yielded before sufficient pressure could be brought to bear on pyramid *a* to shear off the fragment *c* still adhering to pyramid *b*.

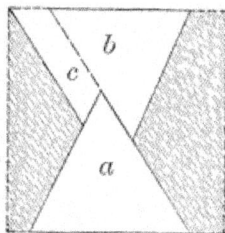

When only one pyramid was formed it was generally well developed, and in some cases its apex reached nearly to the opposite bed-face.

The production of but one pyramid is perhaps an indication of a peculiar structural condition of the stone, combined with approximate parallelism of the end-faces of the cube and a proper uniform bearing of the latter against the pressing-plates of the testing-machine. If the substance cementing together the quartz particles of the material is rather more indurated at one end of the specimen than at the other, the molecular motion induced by the pressure will be more pronounced at the latter end, and the formation of an opposing pyramid be prevented. Mr. Rennie mentions as "a curious fact in the rupture of amorphous stones, that pyramids are formed, having for their base the upper side of the cube next the lever, the action of which displaces the sides of the cubes, precisely as if a wedge had operated between them." Mr. Clark says, concerning sandstones, that "after fracture the upper portion generally retained the form of an inverted square pyramid, very symmetrical, the sides bulging away in pieces all round."

The conclusion derived from the above quotations, that the base of a solitary pyramid is generally found next the moving or driving head of the press, was not entirely corroborated in the Watertown experiments, although the phenomenon seems to occur more frequently at that end than at the opposite one. The assumption of a slight decrease or increase in the strength of the cementing material from one end of the cube to the other would go far to explain the matter. There exist also many gradations from the formation of a large isolated pyramid

2

to that of two smaller but well-developed pyramids. Frequently one of the two pyramids preponderated in size and regularity of form, while the other was only rudimentary.

Without exception, the Haverstraw freestone yielded either suddenly, without previous warning, or the first crack or other evidences of destructive strain appeared only when the ultimate load had been nearly reached. All cubes, and more especially those from six inches on a side upwards, burst with a dull explosive sound.

Of the several varieties of material experimented upon at the Watertown Arsenal, the samples of freestone were the last to be tested, as they were considered to be the most important. They represented the only species of natural stone provided; and in crushing them and drawing deductions from the results it was thought advisable to utilize the information obtained in testing samples of artificial building material. With the latter there is always more or less doubt as to the relative condition of large and small cubes of the same kind. It is quite probable that a 1-inch or 2-inch cube of such material will season sooner than an 8-inch or 12-inch cube. With every additional inch of a cube it is reasonable to assume that its age ought to be increased to render its actual condition similar to that of a smaller cube. Moreover, the amount of labor to be expended in moulding different sizes of cubes or prisms to consolidate them equally requires a nicety of adjustment not attainable in practice.

This difficulty does not exist with quarried natural stone. If all of the samples are taken from the same part of the quarry, and treated exactly alike, it is to be presumed that the results of the tests are fairly comparable.

PREPARATION OF BED-FACES OF SPECIMENS.

In order to develop the full strength of the stone it was necessary to decide upon a method of finishing the beds of the samples, so as to insure a uniform bearing against the smooth holder-plates of the machine.

The cubes ranged by increments of an inch from one inch

to twelve inches on a side. There were four samples of each set, except the 1-inch set, of which there were only two.

The 2-inch, 3-inch, 4-inch, and 5-inch sets were selected for making preliminary comparative tests. Two samples of each of these sizes were once more carefully rubbed with water and fine sand upon a smooth iron plate until their beds were as smooth and plane as it was possible to make them. The other four pairs were simply plastered, the slight unevenness of their faces being covered and smoothed off by a film of plaster of Paris.

The following table, corrected from General Table I. for observed pressure per square inch of bed-surface, shows the results of these comparative tests:

TABLE A.

CRUSHING RESISTANCE OF CUBES OF HAVERSTRAW FREESTONE WITH THEIR
BED-FACES FINISHED BY EXTRA RUBBING AND BY PLASTERING.

Size.	RUBBED BEDS.		PLASTERED BEDS.	
	Strength of Cube.	Average Strength.	Strength of Cube.	Average Strength.
2-inch Cube...............	23,816 lbs.	23,402 lbs.	26,856 lbs.	24,602 lbs.
2-inch Cube...............	22,988 lbs.		22,348 lbs.	
3-inch Cube...............	55,638 lbs.	53,914 lbs.	64,818 lbs.	58,486 lbs.
3-inch Cube...............	52,191 lbs.		52,155 lbs.	
4-inch Cube...............	101,456 lbs.	93,448 lbs.	95,200 lbs.	97,304 lbs.
4-inch Cube...............	85,440 lbs.		99,408 lbs.	
5-inch Cube...............	141,450 lbs.	136,987 lbs.	201,300 lbs.	186,000 lbs.
5-inch Cube...............	132,525 lbs.		170,700 lbs.	

The results exhibited in this table indicated that it would be safe to plaster the bed-faces of the remaining cubes as well as those of the prismatic slabs of freestone. This economical and convenient mode of preparing stone samples for compressive tests appears to be trustworthy when the beds have been previously rendered as smooth and true as possible by hammer, chisel, and by rubbing, and when the film of plaster is as thin as possible.

The tests were carried as far as the capacity of the machine permitted. Three of the 12-inch cubes resisted the maximum load of 800,000 pounds; they were subsequently tested combined as a pier. One of the 10-inch cubes exhibited unexpected strength as compared with other cubes of the same size; it was not broken under the maximum load, while the weakest stone of that set failed under a pressure of 521,000 pounds.

The average resistance of 9-inch cubes per square inch of surface under pressure varied from 5494 to 7886 pounds. There was not much difference in strength between the individual samples in the sets of 6-inch, 7-inch, and 8-inch cubes, respectively; but the average strength of the 6-inch cubes considerably exceeded that of the other two sets named. The highest average resistance per square inch of bed-surface was obtained with the 1-inch, 5-inch, and 6-inch cubes, being over 7000 pounds; the mean strength per square inch of bed-surface of the 2-inch, 3-inch, and 4-inch cubes was 6150, 6498, and 6081 pounds, respectively. These data refer to cubes whose beds had been plastered for uniformity of comparison.

The variations in the amount of resistance per square inch of bed-surface developed by individual cubes of each set, and what is more important, between the various sets themselves, show the necessity of a great number of tests to secure a sufficiently reliable estimate of the average strength of freestone, and probably of any other variety of building stone.

COMPRESSIVE RESISTANCE OF VARIOUS-SIZED CUBES.

The experiments which form the subject of this report afford data for a further study of the question of the truth of the empirical law derived from former tests made on a small scale, according to which the resistance per square inch of bed-surface of cubes increases in a certain ratio with an increase of their sides.

In that part of my report of August 10, 1875, in which I discussed the subject of apparent increase of strength of cubes per square inch of bed-surface as the cubes increase in size, it was stated that for cubes of the small size tested it appears that, "if certain cubes of unit dimensions are built together, with cement equal to their own substance, into a cube of larger dimensions and of homogeneous strength, the resistance to compression per square inch of bed-surface increases as the half-ordinates of a cubic parabola."

The equation given for the curve was

$$y = a \times \sqrt[3]{\overline{x}},$$

in which a = average pressure in pounds required to crush a 1-inch cube;

y = pressure in pounds per square inch of bed that would crush a cube the side of which measures x inches.

This empirical law was based upon two series of tests. One series comprised cubes of yellowish-gray Berea stone, increasing by increments of $\frac{1}{4}$ of an inch, from $\frac{1}{4}$ of an inch to 3 inches on a side, with the addition of a single 4-inch cube, all crushed between wooden cushion-blocks. The other series consisted of cubes of bluish Berea stone from 1 inch to $2\frac{3}{4}$ inches on a side, broken between steel plates.

The curve-diagrams constructed from the average results of these tests show a very close approximation to the requirements of the law, excepting only the $2\frac{1}{2}$-inch cubes of the second series.

It was further stated, that it is doubtful whether this law continues up to the ordinary dimensions of building blocks, and that it was not borne out by experiments made in the Brooklyn Navy Yard with a 2000-ton press, by which five 11-inch cubes of Berea stone were crushed. The report went on to say, " Whether the action of these stones [the 11-inch Berea cubes] was anomalous from specific causes, or whether from general causes the law of the increase of strength per square inch fails at a particular value of x, it is impossible to say positively without additional trials. But these large stones broke invariably by splitting vertically in large flakes or sheets, varying from 2 inches to $\frac{1}{4}$ of an inch in thickness, and quite regular over the greatest part of their surfaces of fracture, especially the thinner ones. It is by no means impossible that all rocks have, more or less, a series of joints, somewhat resembling slaty cleavage, along which they open more easily than in any other direction. . . . They [the 11-inch cubes] crushed at somewhat less recorded resistance per square inch of bed than 2-inch cubes of the same stone."

The recent tests at the Watertown Arsenal also failed to show the continuance of this law beyond small cubes.

There is not much information in published works on the compressive strength of stone cubes of various sizes. The following table gives some results obtained by foreign experimenters:

TABLE B.

COMPRESSIVE STRENGTH OF CUBES OF BRITISH BUILDING-STONE.

KIND OF STONE.	Length of Side of Cube. Inches.	Crushing Weight per square inch. Gross Tons.	AUTHORITY.
Aberdeen blue granite......	1	3.47	Vicat.
Aberdeen blue granite......	1½	4.87	Rennie.
Peterhead granite..........	1	2.80	Vicat.
Peterhead granite..........	1½	3.70	Rennie.
Bramley Fall sandstone....	1	2.50	Vicat.
Bramley Fall sandstone....	1½	2.70	Rennie.
Craigleith sandstone.......	1	1.40	Vicat.
Craigleith sandstone.......	1½	2.45	Rennie.
Craigleith sandstone.......	2	3.50	{ Commissioners on stone for Houses of Parliament.
White statuary marble	1	1.43	Rennie.
White statuary marble	1½	2.70	Rennie.
Portland limestone..........	1½	2.03	Rennie.
Portland limestone..........	2	1.66	Rennie.
Portland limestone..........	1	1.17	Vicat.
Portland limestone.........	2	1.50	Institute British Architects.
Portland limestone.........	2	1.74	{ Commissioners on stone for Houses of Parliament.
Bath (Box) limestone.......	1	0.54	Vicat.
Bath (Box) limestone.......	2	0.66	{ Commissioners on stone for Houses of Parliament.

This table shows that the experiments were confined to

small cubes; that except in one case the strength of different sizes of cubes of apparently the same kind of stone was determined by different parties; and that in all cases but one (Portland limestone by Rennie) the larger cube is decidedly stronger per square inch of surface under compression than the smaller one of the same kind. The ratio of increase of strength varies, however, with the several classes of stone. With some varieties, viz., Bramley Fall sandstone, statuary marble, Portland limestone (referring to the tests by Vicat and by the Commissioners on stone for the Houses of Parliament, respectively), and Bath limestone, the increase is approximately in conformity to the cubic formula given in my former report. The observed strength of Aberdeen granite is about 10 per cent lower than required by the formula, while that of Peterhead granite is 13.3 per cent greater. The actual strength of the $1\frac{1}{2}$-inch and 2-inch cubes of Craigleith sandstone, as compared with that of the 1-inch cube, is about 35 and 50 per cent, respectively, in excess of their computed strengths.

Again, according to Barlow, Portland stone crushes at from 1384 to 4000 pounds per square inch; but in the experiments by the Royal Institute of British Architects (1864) the mean resistance to crushing, per square inch, was, for 2-inch cubes, 2576 pounds; for 4-inch cubes, 4099 pounds; and for 6-inch cubes, 4300 pounds. These experiments show an increase in strength of the 4-inch over the 2-inch cubes, in the ratio of the cube root of the square of the side instead of the cube root of the side, as in the Staten Island formula; the strength of the 6-inch cube, compared with that of the 2-inch cube, increased about in the proportion of the square root of the side.

Rondelet, according to Hodgkinson, found that cubes of malleable iron and prisms of various kinds of stone were crushed under loads which varied directly as their areas. Rennie's experiments with cast-iron and wood make it appear that the resistance, particularly in wood, increases in a higher ratio than the area.

In an article in *The Builder*, 1872, the writer says that, "with regard to the supposition that the crushing strength of stone increases with the size of blocks, there has yet been too little proof put forward on which to lay down any law. In fact, the few experiments made by Mr. Kirkaldy bearing on this subject, some of the results of which have been placed at my disposal, go to prove that there is no increase in the resistance to crushing, consequent upon increase in the size of the blocks."

The average strength of 1-inch cubes of Haverstraw freestone tested at the Watertown Arsenal was 7030 pounds per square inch. This was exceeded by the 5-inch and 6-inch cubes, which yielded under average pressures of 7440 and 7354 pounds, respectively. According to the law deduced from the Staten Island experiments, we have

$$y = a \sqrt[3]{x} = a \times x^{0.333};$$

but actually we have for 5-inch freestone cubes, $y = a \times x^{0.085}$; for 6-inch cubes, $y = a \times x^{0.075}$; a being $= 7030$ pounds.

On the supposition that the two 1-inch cubes were of exceptional strength, and taking the 2-inch cubes, the average strength of which was 6150 pounds per square inch, as a basis for comparison, we obtain results approaching more nearly to the formula. The value of a would then, of course, be reduced. In this case we have for the average of the 5-inch cubes $y = a \times x^{0.21}$, and for the strongest of the two (8052 pounds per square inch) as much as $y = a \times x^{0.5}$, or $a \sqrt{x}$. For the 6-inch cubes (average 7354 pounds) we get $y = a \times x^{0.1}$. The strongest of the 10-inch cubes could not be crushed under the maximum load of 800,000 pounds, but a slight seam was opened along one corner. Assuming that the piece might have yielded under a pressure of 840,000 pounds, its crushing load would have been 8400 pounds per square inch, which, as compared with the 2-inch cube, would be equivalent to

$$y = a \times \sqrt[5]{x} = a \cdot x^{0.2}.$$

When it is considered that the experiments at Staten Island, on which the law of increasing resistance with increasing size of cubes is based, were conducted with the greatest care, it may well be asked why the rule which has been proved to be applicable to a series of small cubes of Berea sandstone either actually fails or only partially and incompletely applies to the larger cubes. The answer to this question is implied in the quotation already made from the former report.

In preparing small cubes for the tests, the soundest pieces are necessarily selected; any material in which flaws, hair cracks, or any other deficiencies can be detected on careful examination, is rejected. The test-piece is naturally designed to be a perfectly sound specimen of its class. Within rather narrow limits, it is possible that, owing to such careful selection, pieces of the same kind of material but of varying sizes are uniform as to texture and identical in homogeneity, and under such conditions it may be taken for granted that some law approximately applies.

The difficulty of close examination and proper selection increases with the greater size of cubes. The stone appears, perhaps, on the outside, quite sound and of uniform texture, but through its mass it may want homogeneity of structure; the material cementing together the grains may be weak in parts, and the grains themselves of varying strength; and there may be cavities, cracks, and soft patches inside of the mass. These defects can be discovered when a large block is split to cut it into smaller cubes, for which the soundest parts are chosen; but the probability that the specimen contains unsound parts increases with the size. This will also explain the fact that cubes of the same size and kind occasionally vary greatly in strength. The weakest of the 9-inch freestone cubes had 35 per cent less resistance than the strongest; and the weakest of the 10-inch cubes probably fully 60 per cent less than the strongest of that kind.

In practice, a comparatively large cube ceases to be a unit, but is rather a conglomerate of smaller irregular pieces, joined together by a cementing substance of varied strength, and perhaps partially separated by minute cracks, cavities, or pores. Under such conditions the stone cannot develop the same strength as if it were a true unit.

In other words, according to the quotation referred to, cubes of certain unit dimensions may be conceived to be built together with cement equal in strength to their own substance, into a cube of greater size, producing a true monolith of homogeneous structure and corresponding strength.

Judging from the tests made with small cubes of Berea stone, we should expect the resistance to compression per square inch of bed-surface of a true monolith to materially increase with its size. Even assuming the masses of which an actual specimen is built up to be of uniform strength, especially when of the quartzose variety, it is probable that the cementing substance, whether silica, carbonate of lime or magnesia, oxide of iron, alumina, or mixtures of one or more of them, is of variable strength and density in different parts of the stone; its adhesion to the parts it binds together may be less perfect at some places than at others; and the actual ultimate resistance of an apparent monolith will then be less than the calculated one. As the loading progresses, incipient cracks, quite imperceptible to the observer, will be formed where the cementing substance is weakest, and seams of more or less extent will open, much as in brickwork under pressure. With brittle material like freestone, the very jar of sudden internal yielding will act like a blow on adjacent parts, weaken the cohesion of the cement in the vicinity and its adhesion to the unit particles it binds together, and further yielding will ensue. If these initial, though inappreciable, cracks run about parallel to the bed, the aggregate cube ceases to be a monolith; and it is known and has been again proved by tests made in that direction at the Watertown Arsenal, that a cube built up in several courses is inferior in strength to a solid cube. The conditions are more unfavorable when, owing to defective strength of the cementing substance, initial cracks open approximately parallel to the line of pressure; the stone will then be divided into irregular columns, the heights of which may considerably exceed the least dimension of their cross-section, inducing transverse bending or bulging, and premature separation of parts by cleavage and splintering off. It is more probable, however, that early partial yielding occurs in a more complicated manner, or in various oblique directions through the mass, which will still more favor disintegration under a comparatively moderate pressure. In former experiments at Staten Island several samples of sandstone, in the form of 2-inch cubes, displayed greater strength when broken on edge than when crushed on bed. It may be inferred from this that the cubes broken on bed had weak cement joints in a direction normal to the bed, favoring lateral cleavage; and that this kind of defect either did not exist, or was at all events of much less consequence, when the cube was broken on edge. It is possible that the clamping action of the holder-plates between which the test-piece is held is reduced in its effect as the distance between them increases. A flaw in a 2-inch cube favoring an incipient crack through its central part will not affect the strength to such an extent (from the nearness of the friction-plates) as cracks tending to separate laterally pieces of similar or even greater thickness from a larger cube.

Perfect homogeneity of structure is necessary to develop the full strength of stone or similar material. That Haverstraw freestone is deficient therein, is shown in the strain-diagram to be referred to hereafter.

We may safely conclude that those cubes which exhibited the greatest resistance in their class approached most nearly the state of comparatively perfect condition. We further believe that the law, perhaps more or less modified, would be corroborated if it were possible to provide a series of cubes of varying sizes, each of which was truly homogeneous throughout.

Berea sandstone evidently possesses a remarkable degree of homogeneity of structure, at least up to cubes of 3 or 4 inches on a side; and it is quite possible that if it had been tried in larger pieces, the results would have been approximately in conformity to the empirical law. It failed, however, with 11-inch cubes, as already stated; and might have done so with somewhat inferior sizes.

With artificial stone, like cement, mortar, and concrete, all of which were consolidated by ramming or tamping in moulds, another element enters the question which influences the strength of the piece. A certain amount of labor in ramming or beating is performed in making, for instance, a 1-inch cube. How much work should be applied in consolidating a 2-inch,

6-inch, or 12-inch cube? It is known that, within certain limits, repeated rolling of a wrought-iron bar with accompanying reduction of cross-section increases its homogeneity and strength, while it also renders it more brittle. It is probable that a certain amount of ramming, with a corresponding weight of the ramming tool, may render a large cube as homogeneous through its entire mass as a reduced amount of work usually expended upon a smaller cube, but the law of this proportion is not known.

The faces of some of the larger cubes of neat cement, previous to being tested, exhibited numerous minute hair-cracks, crossing each other in all directions, but distinguishable only after moistening the surface. This sort of examination was limited to a few samples; it was presumed that the rest would not differ in that respect. The cracks were evidently due to irregular shrinkage while the cement was setting and hardening. This process naturally went on quicker in the outer crust than in the core of the cube; in hardening, the contraction of the outer portions was more or less obstructed by the inner mass which had not so far advanced in setting and change of volume. To all appearances the cubes of neat cement were entirely sound and in good condition; but it is not doubted that these incipient cracks, which must have extended for some depth into the mass of the cube, impaired its strength. In this respect, therefore, the small cubes ought to have been—as they really were—proportionally stronger than the larger ones, since the hardening or seasoning from the shell to the centre must have been quicker, more complete, and more uniform.

There is no reason to doubt that the cubes of mortar or concrete, which had been moulded in precisely the same manner as the samples of neat cement, would have shown similar hair-cracks caused by shrinkage if their rough exterior had not prevented their being distinguished.

The fact that the cubes of cement, etc., were not kept immersed in water, but only covered up with sand, may to some extent account for irregularities in the results. Mr. Whitaker, who conducted numerous experiments for Mr. Grant on behalf of the British Government, found that 12-inch concrete cubes, rammed into moulds by hand-beating with a mallet, resisted under compression an average of 30 per cent more than concrete cubes of the same size made in the ordinary way; he also found that 12-inch cubes set in water for one year stood a greater weight than those set in air during the same period, while 6-inch cubes were stronger set in air than in water.

We infer from the Watertown experiments that with material lacking homogeneity of structure the strength of cubes is not as great as required under the law, although significant traces of its applicability may be discovered with pieces which exhibited superior resistance. The question still remains unsettled whether stone, approximately homogeneous, when in the form of larger blocks or cubes exhibits greater compressive strength per square inch of bed-surface than smaller cubes. It would seem to be desirable to continue experiments with the same kind of Berea stone that furnished the data on which the law was founded, and to try other species of building-stone which, from preliminary tests, may promise to possess a high degree of homogeneity of structure.

STRENGTH OF SIMPLE AND COMBINED PRISMS OF VARYING HEIGHT.

A number of tests were also made at the Watertown Arsenal in order to ascertain the behavior and relative compressive strength of square prisms of less height than cubes of the same cross-section. Some of the prisms were made of Haverstraw freestone, and others of neat Dyckerhoff cement.

On examining and comparing the results obtained with prisms of varying height, it seemed to be possible to express the law connecting strength and form of specimens by some formula.

Some unit of strength was evidently required to be introduced into such a formula. The law referring to the strength of cubes of varying size having been found to be inapplicable to the specimens, the usual method of assuming a unit pressure per square inch of bed-surface, represented by the arithmetical mean of the average crushing resistances of the several sizes of cubes tested, naturally suggested itself. The series of freestone samples actually broken on the first application of the ultimate pressure within the maximum load of 800,000 pounds embraced cubes from 1 inch to 11 inches on a side, excepting one 10-inch cube. The column of observed loads in the following Table C shows that the arithmetical mean of all the average loads would be 6600 pounds per square inch of bed-surface. But the observed crushing strength of the 1-inch cubes greatly exceeds that of all other sizes, with the exception of the 5-inch and 6-inch cubes; the cubic contents of the individual prisms are, moreover, from sixteen to several hundred times greater than that of a 1-inch cube; and it seems to be, therefore, justi-

TABLE C.

COMPRESSIVE STRENGTH OF CUBES OF HAVERSTRAW FREESTONE.

Side of Cube.	Observed Ultimate Loads, in Pounds.				Computed Load of Cube, in pounds, on the basis of 6550 pounds per square inch.	Excess or Deficiency of Computed Load.
	Of Cubes, Singly.		Averages.			
	Per Square Inch.	Of Whole Cube.	Per Square Inch.	For Whole Cube.		
1 inch......	6,959	6,950	7,030	7,030	6,550	− 7.3%
1 inch......	7,102	7,102				
2 inch......	6,714	26,856	6,150	24,600	26,200	+ 6.1%
2 inch......	5,587	22,348				
3 inch......	7,202	64,818	6,498	58,482	58,950	+ 0.8%
3 inch......	5,795	52,155				
4 inch......	5,930	94,200	6,081	97,296	104,800	+ 7.2%
4 inch......	6,213	99,408				
5 inch......	8,052	201,300	7,440	186,000	163,750	− 13.6%
5 inch......	6,828	170,700				
6 inch......	7,179	258,444	7,354	264,744	235,800	− 12.3%
6 inch......	7,048	253,728				
6 inch......	7,471	268,956				
6 inch......	7,719	277,884				
7 inch......	6,115	319,635	6,156	301,644	320,950	+ 6.0%
7 inch......	5,728	280,673				
7 inch......	6,590	322,910				
7 inch......	6,190	303,310				
8 inch......	6,219	398,016	6,271	401,344	419,200	+ 4.3%
8 inch......	6,674	427,136				
8 inch......	6,040	386,560				
8 inch......	6,152	393,728				
9 inch......	5,769	467,289	6,534	529,254	530,550	∓ 0.2%
9 inch......	6,989	566,109				
9 inch......	7,836	638,764				
9 inch......	5,494	445,014				
10 inch......	5,210	521,000	6,673	667,350	655,000	− 1.9%
10 inch......	6,638	663,800				
10 inch......	8,400*	840,000				
10 inch......	6,446	644,600				
11 inch......	6,508	787,468	6,418	776,578	792,555	+ 2.0%
11 inch......	6,453	780,813				
11 inch......	6,440	779,240				
11 inch......	6,270	758,670				

* This 10-inch cube was not crushed under the available maximum load of 800,000 pounds. In the table it is assumed that it might have yielded under 40,000 pounds of additional pressure.

fiable to omit the smallest set of cubes from the calculation. The average crushing load of the several cubes from 2 to 11 inches on a side is found to be 6550 pounds per square inch, which the following table shows to give quite satisfactory results when the loads thus computed are compared with those actually observed. It should be stated that these observed loads are those only of cubes the beds of which had been plastered so as to render the conditions of fracture uniform.

The greatest differences between computed loads and averages of observed loads are found in the sets of 5-inch and 6-inch cubes, and even there the difference does not reach 14 per cent. It is thought that 6550 pounds, the general average crushing stress per square inch of bed-surface for cubes of Haverstraw freestone, may be considered fairly applicable to prisms of the same kind of material, obtained at the same time from the same part of the quarry, and wrought and tested under precisely the same conditions.

Prisms of Haverstraw Freestone.—Two series of square prisms of less height than cubes of the same cross-section were tested.

One series contained prisms 4″ × 4″ on bed, and 1, 2, and 3 inches in height, respectively. The other series measured 8″ × 8″ on bed, with heights of 2, 3, 4, 5, 6, and 7 inches, respectively. There were two prisms to each set.

It was noticed that the prisms generally gave earlier warning of approaching destruction than the cubes, crackling noises being audible during the later stages of loading. This is probably due to the frictional resistance of the pressing-plates, which, from being nearer together, hold the prisms in a firmer grasp than the cubes, and therefore permit disintegration to proceed without ultimate fracture for a longer period.

The testing-machine did not prove powerful enough to crush either of the two 8″ × 8″ × 2″ prisms: one of them was apparently almost intact when removed, some small spawls only having cracked off from the edges; the other had suffered a little more, but both samples would evidently have resisted considerably more pressure.

In prisms of half the height of corresponding cubes the formation of pyramidal fragments began to be fairly developed, becoming more complete as the height increased. The thinner prisms were simply broken up into numerous small, irregular pieces, besides being to some little extent ground to powder; what core remained could easily be broken up by hand. There were only faint traces of pyramid formation.

It has long been known to close observers that the compressive strength of prisms increases as their height diminishes. Mr. Navier, however, was of the opinion that the force necessary to produce crushing is greatest when the piece has the form of a cube, and diminishes when the piece is lower or higher. Mr. Hodgkinson says on this subject: "Shorter specimens generally bear more than larger ones of the same diameter or dimensions of base. In the shortest specimens fracture takes place by the middle becoming flattened and increased in breadth (bulged), so as to burst the surrounding parts and cause them to be crumbled and broken in pieces. This is usually the case when the lateral dimensions of the prism are large compared with the height."

That such spreading out across the middle part of the prism takes place is shown by the chips and spawls that gradually fly or drop off from the exposed sides of the piece, leaving a rough, irregularly triangular groove around the prism, or merely a rough, slightly concave indentation, as in the case of the $8'' \times 8'' \times 2''$ freestone prisms which could not be broken.

A case slightly analogous to that of short prisms under compressive stress occurs in testing the tensile strength of iron, steel, and other metals. A bar of certain cross-section will develop far more tensile resistance when its exposed length is very small compared to its diameter than when it is several times that dimension. Or, as Mr. Kirkaldy deduced from his experiments, "the breaking strain is materially affected by the shape of the specimen. The amount borne was much less when the diameter was uniform for some inches of the length than when confined to a small portion—a peculiarity previously unascertained, and not even suspected. It is necessary to know correctly the exact conditions under which any tests are made before we can equitably compare results obtained from different quarters."

Professor Weyrauch, referring to the above observations, says that the stress for compression should show a similar difference, and that this, according to Bauschinger and others, is found to be the case.

While the fact of an increase of compressive resistance with a diminution of the height of prism was more or less known, no attempt seems to have been made to determine the probable ratio of such increase when the height of the prism becomes less than that of a cube.

In endeavoring to arrive at an empirical law expressing the compressive strength of a prismatic slab, it was considered that as the height of the piece is decreased, the area of bed-surface remaining unchanged, the exposed lateral area becomes smaller, and the liability of the material to be forced out sideways under the internal strain becomes less; due weight must therefore be given in a formula to this relation. Besides assuming some general or uniform crushing load per square inch of bed-surface, representing the average obtained from a series of actual tests, it seemed necessary to introduce into the formula an expression of the relation between areas of bed and sides; of the difference between the heights of cube and corresponding prism; and of the strength of a cube, the area of whose bed is equal to that of the prism.

The following formula is given:

$$W = C + 2m \times (h - h_1)^2 \times \sqrt{p};$$

in which $W =$ crushing load of prism, in pounds;

$C =$ crushing load of a cube having the same area of bed as the prism;

$m =$ crushing load of material per square inch; an average derived from testing a series of cubes of various sizes, and of the same material as the prism;

$p =$ quotient obtained by dividing the area of the bed by the sum of the areas of the sides of the prism;

$h =$ height of cube of crushing strength C, in inches;

$h_1 =$ height of prism, in inches.

For Haverstraw freestone, the value of m would be 6550 pounds, in accordance with preceding explanations and table.

The crushing loads obtained by this formula are compared with the results actually obtained with freestone prisms in Table D, in which the beds of prisms are assumed to be true squares. As such, their bed-areas are very slightly different from those of the prisms actually tested; for which reason the total crushing loads, which are in the table stated to be derived from experiment, necessarily vary a little from those given in General Table I.

TABLE D.

COMPRESSIVE STRENGTH OF PRISMS OF HAVERSTRAW FREESTONE.

SIZE AND MARK OF PRISM.	OBSERVED ULTIMATE OR CRUSHING LOAD IN POUNDS. Of Sample.	Average.	Computed Crushing Load in pounds. $m = 6,550$ lbs.	Excess or Deficiency of Computed Load.
4″ × 4″ × 3″, *a*	98,256	106,856	112,363	+ 5.1%
4″ × 4″ × 3″, *b*	115,456			
4″ × 4″ × 2″, *a*	131,536	128,448	141,852	+ 10.4%
4″ × 4″ × 2″, *b*	125,360			
4″ × 4″ × 1″, *a*	300,544	262,840	222,700	− 15.3%
4″ × 4″ × 1″, *b*	225,136			
8″ × 8″ × 7″, *a*	428,096	423,232	426,200	+ 0.7%
8″ × 8″ × 7″, *b*	418,368			
8″ × 8″ × 6″, *a*	401,984	418,208	449,452	+ 7.4%
8″ × 8″ × 6″, *b*	434,432			
8″ × 8″ × 5″, *a*	444,268	497,036	493,765	− 0.7%
8″ × 8″ × 5″, *b*	549,804			
8″ × 8″ × 4″, *a*	597,504	547,264	567,408	+ 3.6%
8″ × 8″ × 4″, *b*	497,024			
8″ × 8″ × 3″, *a*	656,064	610,368	686,600	+ 12.5%
8″ × 8″ × 3″, *b*	564,672			
8″ × 8″ × 2″, *a*	Not broken by maximum load of 800,000 lbs.	800,000+	890,800	?
8″ × 8″ × 2″, *b*				

Examining the table, it is seen that material divergence between observed and computed loads occurs only in the case of the 4″ × 4″ × 1″ prisms, the difference being 15.3 per cent. This may perhaps be accounted for by the difficulty of determining with precision when a very thin prism has really given way, because with such specimens the moment of absolute yielding is by no means as distinctly marked as with thicker prisms.

The falling off in observed average strength of the $8'' \times 8'' \times 6''$ prisms, when compared with the preceding set of 7 inches in height, is probably due to some structural defect in the block from which these prisms were cut.

On the first pages of this report it is stated that from previous tests the average strength of a prism of blue Berea sandstone, 2 inches square and 1 inch in height, crushed between steel, had been found to be 75,888 pounds. In the report of August 10, 1875, on the compressive strength, etc., of buildingstone, Table IV. gives the strength of eight 2-inch cubes of that material. Excluding one specimen on account of excessive weakness,—it being about 40 per cent less in strength than the average of the others,—the mean resistance of the seven remaining cubes is 51,671 pounds, or 12,918 pounds per square inch.

For nearly homogeneous stone, as blue Berea stone as far as tested appears to be, the prismatic formula would have to be modified, inasmuch as the value of m becomes variable, i.e., m will be $= a \times \sqrt[3]{h}$, in which $a =$ pressure in pounds needed to crush an inch cube and $h =$ side or height of cube in inches. The load C, of a cube having the same area of bed as the prism, would be

$$C = h^2 \times a \sqrt[3]{h} = a \times h^{2.333};$$

and the formula in its modified form,

$$W = ah^{2.333} + 2a\sqrt[3]{h} \times (h - h_1)^2 \times \sqrt{p}$$
$$= a \times \{h^{2.333} + 2\sqrt[3]{h} \times (h - h_1)^2 \times \sqrt{p}\}.$$

Referring to the $2'' \times 2'' \times 1''$ prisms of Berea stone, we have

$$a = 10,252 \left(= \frac{12,918}{\sqrt[3]{2}} \right) \text{ pounds};$$

$$h = 2 \text{ inches};$$
$$h_1 = 1 \text{ inch};$$
$$p = .5;$$

therefore

$$W = 10,252 \times \{2^{2.333} \times 2\sqrt[3]{2} \times 1^2 \times \sqrt{.5}\} = 69,937 \text{ pounds,}$$

or 7.8 per cent less than the average of the observed loads of
seven prisms, but higher than two of the latter.   The record of
another set of four tests of blue Berea sandstone prisms, each
$2'' \times 2'' \times 1''$, crushed under steel, likewise given in Table IV.
of the former report, shows an average resistance of 69,550
pounds per sample—almost identical with the computed load.

Prisms of Neat Portland Cement.—The greater portion
of the cement cubes were broken directly between the steel
and gun-iron plates of the machine, while the balance of the
cubes, and all of the prisms, had their beds previously plastered.   This, and the fact that there was more or less divergence of ultimate resistance among samples of the same set of
cubes and among the various sets of different sizes, renders it
somewhat difficult to fix upon a suitable value of an average
crushing resistance per square inch, to be introduced as coefficient m in the prismatic formula.

The average ultimate crushing strength of six 1-inch cement cubes was 5896 pounds per square inch.   The average
resistance of the six 2-inch cubes was 7094 pounds per square
inch: nearly the proportion, as compared with the 1-inch
cube, required under the cubic formula.   The resistance per
square inch of the following sizes is not in conformity to that
law, however.   The 3-inch cubes broke under an average load
of but a few pounds more than the 1-inch cubes, and the
averages of all the larger cubes, from 4 to 11 inches on a side,
varied from 4283 to 5374 pounds.   To decide upon a general
average compressive resistance per square inch, corresponding
to m in the prismatic formula, the aggregate ultimate resistance of all of the cubes from 1 inch to 11 inches on a side (the
12-inch cubes being excluded, as some of them were not
broken under the first application of the maximum load),
amounting to 15,065,604 pounds, was divided by 3036, the
aggregate area in square inches of the bed-surfaces of these
cubes, giving a quotient of 4962 pounds.   As there was some
uncertainty as to the accuracy of this value, the round number
5000 pounds was adopted as representing approximately the
average strength per square inch of Dyckerhoff cement, that is,
the new value of m in the prismatic formula.   A comparative

table of ultimate resistances of cubes, giving the loads computed on the basis of 5000 pounds per square inch and the several observed loads and their averages, is found in the part of this report relating to cement ; it will be seen that there is a tolerably fair agreement among them, except with the smallest sizes of cubes. The samples when tested were from 22 to 23 months old.

Applying the prismatic formula to cement, Table E results, in which the usual correction is made, from General Table II.

TABLE E.

COMPRESSIVE STRENGTH OF PRISMS OF NEAT DYCKERHOFF'S PORTLAND CEMENT.

SIZE AND MARK OF PRISM.	OBSERVED ULTIMATE OR CRUSHING LOAD, IN POUNDS.		Computed Crushing Load in Pounds. $m = 5,000$ lbs.	Excess or Deficiency of Computed Load.
	Of Samples.	Average.		
4″ × 4″ × 3″, *a*	85,712			
4″ × 4″ × 3″, *b*	96,080	96,043	85,775	− 10.69%
4″ × 4″ × 3″, *c*	106,336			
4″ × 4″ × 2″, *a*	101,760			
4″ × 4″ × 2″, *b*	101,344	101,920	108,284	+ 6.21%
4″ × 4″ × 2″, *c*	102,656			
4″ × 4″ × 1″, *a*	242,112			
4″ × 4″ × 1″, *b*	268,912	261,104	170,000	− 34.89%
4″ × 4″ × 1″, *c*	272,312			
8″ × 8″ × 6″, *a*	341,056			
8″ × 8″ × 6″, *b*	392,192	369,344	343,100	− 7.1%
8″ × 8″ × 6″, *c*	374,784			
8″ × 8″ × 5″, *a*	419,968			
8″ × 8″ × 5″, *b*	374,016	385,271	376,925	+ 2.2%
8″ × 8″ × 5″, *c*	361,828			
8″ × 8″ × 4″, *a*	389,889			
8″ × 8″ × 4″, *b*	387,200	380,928	433,136	+ 13.7%
8″ × 8″ × 4″, *c*	365,690			
8″ × 8″ × 3″, *a*	566,824			
8″ × 8″ × 3″, *b*	413,888	460,237	524,125	+ 13.9%
8′ × 8″ × 3″, *c*	400,000			
8″ × 8″ × 2″, *a*	642,688	682,496	680,000	− 0.36%
8″ × 8″ × 2″, *b*	722,304			
8″ × 4″ .76 × 2″	317,500	317,500	340,000	+ 7.2%
12″ × 12″ × 8″, ...	Not		818,000	
12″ × 12″ × 6″, ...	broken		974,556	
12″ × 12″ × 4″, ...	singly.		1,274,772	
12″ × 12″ × 2″, ...			1,944,750	

of Cement Tests, for fixing the total observed pressure on prisms with truly square beds.

In the foregoing table the greatest divergence between observed and computed loads is again found in the thinnest set, or the $4'' \times 4'' \times 1''$ prisms, most probably for the same reason as suggested in the case of freestone prisms of similar size. Numerous slight crackling sounds were heard while testing these thin slabs long before the moment when it was concluded that the ultimate load had been reached ; when removed, the piece was found to be well disintegrated. The large straining-plates of the machine being for such samples only one inch apart, a close observation of their behavior under stress is not practicable, and in assuming a certain load as the actual crushing strength large allowance must be made for personal equation. In Mr. Grant's tests of the crushing resistance of cement bricks, it is reported that each specimen showed signs of giving way with considerably less pressure than that which finally destroyed it, the ratio of the weight which produced the first crack to that which finally crushed it being nearly 5 to 8. While testing the $4'' \times 4'' \times 1''$ cement prisms of the preceding table, crackling sounds began to be heard under a load of 140,000 pounds in piece a, of 40,000 pounds in piece b, and of 100,000 pounds in piece c. The $8'' \times 8'' \times 2''$ prisms (for which p has the same value as for the smaller prisms just referred to) exhibit a remarkable coincidence of observed and computed loads. The straining-plates being twice as far apart for the larger prisms, better facilities for observation existed.

The table shows that the prismatic formula was also tried with a slab not square, but rectangular in cross-section ; it measured $8'' \times 4''.76$ on bed, with 2 inches height. The piece had originally been an $8'' \times 8'' \times 2''$ prism, but while being put into the press it was accidentally dropped, breaking into three fragments. The largest of these was then carefully trimmed into the form stated, in a shaping-machine at the arsenal. The area of its bed was now 38.08 square inches ; the side of a corresponding cube would therefore be 6.17 inches. With an average strength of 5000 pounds per square inch, adopted for the cement cubes from 3 inches upwards, the total crushing strength of a 6.17-inch cube would average 190,400 pounds. The value of p is 0.7461 $\left(= \dfrac{38.08}{51.04} \right)$; and $h - h_1 =$ 4.17 $(= 6.17 - 2)$; therefore

$$W = 190,400 + 2 \times 5000 \times \sqrt{.74.61} \times 4.17^2 = 340,000 \text{ pounds.}$$

This computed load is found to be only 7.2 per cent in excess of the observed load.

Mr. Reid, in his treatise on cement, says that a brick made of neat Portland cement, nine months old, measuring $9'' \times 4\frac{1}{4}''$ $\times 2\frac{3}{4}''$, and therefore of an area of bed equal to 38.25 square inches, was crushed under a load of 7027 pounds per square inch. According to the empirical rule, the cube corresponding to such a prism would have a length of side of 6.18 inches; p = 0.5239; $h - h_1 =$ 3.43; and the value of coefficient m would in this case be 4871 pounds.

The resistance of a prism increasing as its height diminishes, it may therefore be conceived that it is finally reduced to a film of infinite tenuity, in which condition it can undergo no further deformation even under an immeasurably great pressure. This hypothetical condition is fulfilled by the formula, because h_1 will then be $= 0$, and $p = \dfrac{1}{0} = \infty$; therefore

$$W = C + 2m \times h^2 \times \infty = \infty.$$

To what extent the formula may stand the test of further experiments, especially with other forms of prisms than those described, remains to be seen. It would be desirable to make further investigations for that purpose.

It is possible that with certain modifications the formula can be made to express the average resistance of prisms exceeding the height of a cube. Its applicability in that direction will most probably be limited, however, since the tendency to lateral flexure will have to be considered when the prism attains a certain height. One or another of existing formulas for calculating the strength of cast-iron pillars, suitably modified for stone, may perhaps be arranged to answer in such cases.

Remarks on Prisms higher than a Cube.—There was

but one experiment made in that direction, with a small prism of freestone, 1 inch square in cross-section and 2 inches high. It broke under a load of 4550 pounds—about 77 per cent of the average crushing load (5896 pounds) of a 1-inch cube. The fracture revealed a little pyramid at one end which had apparently acted as a wedge, forcing out the bulk of the piece in the form of three longitudinal fragments, each nearly of the whole length of the prism.

Tests of blue Berea sandstone, made in 1875, show the average proportion of compressive strength between a 2-inch cube and a prism of twice the height of a cube to be as 100 to 89.5.

Mr. Navier gives data from Rondelet to show diminution of strength when the height is greater than side of base. The cross-section of the prisms was square, measuring 5 centimetres or 1.968 inches on a side, equal to 3.875 square inches of bed-surface. The prisms of each set were one, two, and three cubes in height, respectively. The results are shown in the following table, the crushing loads being expressed in pounds:

TABLE F.

COMPRESSIVE STRENGTH OF FRENCH BUILDING-STONE. CROSS-SECTION OF PRISM, SQUARE; HEIGHT, VARIABLE; AREA OF BED, 3875 SQUARE -INCHES. (FROM RONDELET.)

KIND OF STONE.	Height of Prism.	Specific Gravity.	Crushing Load, in Pounds.	Percentage of Strength. Cube=100
a. Lias limestone, very hard............	1 cube	2.388	19,512	100.0
	2 cubes	2.388	11,930	61.0
	3 cubes	2.388	10,538	54.0
b. Hard Stone, Fond de Bagneux...........	1 cube	2.255	14,661	100.0
	2 cubes	2.255	9,315	63.5
	3 cubes	2.255	8,576	58.5
c. Hard Rock, De Chatillon...............	1 cube	2.342	11,328	100.0
	2 cubes	2.342	8,841	78.0
	3 cubes	2.342	8,495	75.0
d. Hard Rock, De Chatillon...............	1 cube	2.199	8,203	100.0
	2 cubes	2.199	6,563	80.0
	3 cubes	2.199	6,372	77.7
e. Hard Rock, De Chatillon.	1 cube	2.162	7,798	100.0
	2 cubes	2.162	6,237	80.0
	3 cubes	2.162	6,067	77.8

With the two lightest and softest sets of prisms the relative diminution of strength as the height of the piece increases is the same, and is less than in the other three sets. The hardest and at the same time the heaviest stone (*a*) suffers the greatest reduction of strength by increasing the height of prism, and the next strongest (*b*) very nearly the same. Set *c*, of medium strength per cube, shows also a medium decline of resistance with increasing height, compared with the softer and harder varieties.

Further experiments on an extensive scale are required to formulate even an approximate law on this subject,—a law which apparently must consider for different kinds of stone, their relative hardness or specific gravity, or both.

Remarks on Prisms Divided in Courses.—Some compound prisms formed of pieces that could not be broken singly were tested.

The three 12-inch freestone cubes which had, each, resisted the maximum load of 800,000 pounds were combined as a pier with dry joints, and were tested in that form.

When this pile had been clamped in the press it was found that the plastered beds which had previously undergone pressure with the single pieces were slightly convex in their middle parts, which prevented a perfectly close joint at the corners, although the gaps at these joints did not exceed the thickness of a sheet of paper. This convexity may possibly be ascribed to the elasticity of the material, which had recovered somewhat more of its original length through the central portion of the cube than at the corners.

The first crack appeared when the load had reached 700,000 pounds, and the pier yielded with a reverberating explosion under an ultimate pressure of 748,000 pounds. It was well shattered, especially the cube next to the straining-press.

Four piers formed of cement prisms 12 inches square on the bed-surfaces were tested, each pier composed of three prisms of the same size.

The pile formed of prisms each only two inches high resisted the maximum load of 800,000 pounds. Each of the other piers, consisting of prisms 4, 6, and 8 inches high, respectively, failed under stresses below the maximum load of the testing-machine.

One of the 10-inch freestone cubes which had proved refractory under the available maximum load, once applied, was subsequently combined into a pier with the three equally refractory cement prisms, each of which measured 12″ × 12″ × 2″. This compound, dry-jointed pier yielded under a stress of 654,000 pounds. At 550,000 pounds the first cracking sound was heard; at 580,000 pounds the prism representing the base of the pier began to flake off at the corners. The pier failed with a loud report, the sides flying off in small pieces; the remaining principal fragments formed two pyramids, that of the freestone being rather sharp-pointed, and reaching nearly to the opposite bed of the cube.

But few records are met with in scientific works on the subject of the strength of building-stone built up in courses.

In Rondelet's "L'Art de bâtir," the strength of prisms of Chatillon rock (specific gravity 2.346), square in cross-section, of 3.875 square inches bed-area, and 3.937 inches height, is given when solid and when divided in courses, as follows:

Prism, in form of a solid body, strength = 11,385 pounds.
Same prism, divided in four courses, " = 9,769 "
" " " " eight " " = 8,153 "

In Stoney's "Theory of Strains" it is said that "Vicat found, from experiments on plaster prisms, that the strength of a monolithic prism whose height is h being represented by unity, we have the strength of prisms:

of 2 courses and of the height, $h = 0.930$;
" 4 " " " $2h = 0.861$;
" 8 " " " $4h = 0.834$;

even without the interposition of mortar. He concludes that the division of a column into courses, each of which is a monolith, with carefully dressed joints and properly bedded in mortar, does not sensibly diminish its resistance to crushing; but he intimates that this does not hold good when the courses are divided by vertical joints."

The curve which can be constructed from the data given by Vicat indicates that there would be little reduction of strength as the number and height of courses increase, which is probably not the case. At all events, there will be a change in the form of the curve when the pier or column is high enough for a development of a tendency to bend transversely, since the ratio of the decrease of strength will then be modified.

The experiments with combined prisms made at the Watertown Arsenal, and by some other investigators, show that stone blocks when arranged or built up in courses have less strength than individual pieces; but while these results are of more or less interest, and will be of use in connection with future similar tests, it is not deemed proper to attempt at present to draw conclusions from a few isolated observations.

It can hardly be said that the cause of loss of compressive strength by dividing a pier into layers or courses without vertical joints is fully understood.

Dupuit is of the opinion that when several prisms bear upon one another, the pressure is unlikely to be transmitted uniformly over the whole surface, and that it may happen, therefore, that some parts will be strained beyond their resistance before a pressure is exerted, which, if uniformly distributed, would have been safely sustained.

This is undoubtedly frequently the case. The bed-faces adjoining each other are never mathematically true and smooth; there are numerous little elevations and depressions distributed all over the surface, which are differently located in the several courses. In some joints the bulk of actual bearing-surface may be in the central portion, in others perhaps rather more toward the margins, and the stress will not pass normally through the mass from top to base. Some courses are also likely to be of less strength than others; when these begin to give way—especially with brittle material—the vibration caused by the sudden destruction of cohesion between parts of one block will react on adjoining courses, intensifying the internal strain to which they are already subjected. By interposing a somewhat elastic cushion in the form of a suitable mortar of sufficient strength, it is probable that the crushing strength of such a pier may be made to exceed that of a dry-jointed pier.   The mortar would improve the defective bearing of adjoining beds, and its elasticity weaken the effect of possibly destructive shocks transmitted from one block to another.

COMPRESSION, SET, ELASTICITY, AND RESILIENCE OF HAVER-STRAW FREESTONE.

[Special Table I. and Strain-sheets I. and II.]

Compression and Set.—Those freestone cubes that measured from 8 inches to 12 inches on a side were tested not only as to their ultimate crushing strength, but also as to rate of compression and amount of set while being loaded.   The results are given numerically in Special Table I., and graphically in Strain-sheets I. and II.

The compression as read off from the micrometer is laid off on the horizontal lines of the sheet.   The length of each large division is equivalent to $\frac{1}{100}$ of an inch; each small division therefore represents $\frac{1}{1000}$ of an inch.   The successive loads applied, as indicated by the scale, are laid off vertically. The height of a large division represents 100,000 pounds; that of a small one, 10,000 pounds.

In the diagrams, the increments of compression and set are therefore the abscissas, and the weights the ordinates.

The observed points of the curve of compression are marked by small black circles.   Where two such circular dots are seen near each other on the same horizontal line, it is understood that the process of loading was here interrupted by relieving the cube from the accumulated pressure, which was then reduced to that initially applied to hold the piece firmly in the machine.   The second dot being to the right of the first shows that some further compression occurred when the load reached the same figure for the second time.

A star at the upper end of a certain curve indicates that the piece yielded and burst while a micrometer observation was being made.

When no star marks the upper end of a curve, it indicates

that the micrometer was there applied for the last time, but that loading was continued until the piece was fractured.

The several small black circles near and parallel to the axis of abscissas show by their distance from the axis of ordinates the amount of set when the load was reduced to the initial pressure. The dotted or broken black lines running from these points up to circular dots of the full-lined curve represent the probable curve of compression under reloading until the pressure before attained is again reached. No observations were taken to determine points of this curve except in the case of a concrete cube, as it would have consumed too much time. It was assumed that renewed compression after the first permanent set had been obtained would proceed more uniformly than at first, because the test-piece had then been more or less relieved from originally existing internal strain.

The initial part of the strain-curve is seen to be always more or less convex toward the horizontal axis, and compression at first proceeds rapidly. Some particles, or molecules of the material, either from comparative inherent weakness, or from not being normally located in reference to others, or from being already overstrained from natural, elementary causes, give way under comparatively small loads. In consequence of this partial yielding, the permanent set observed when the first load of 100,000 pounds is gradually reduced to 5000 pounds is always greater than succeeding increments of set produced by equal increments of load.

The next portion of the curve is approximately straight, or rather is formed of a succession of nearly straight lines of approximately the same angle of inclination, connected by small offsets which mark additional compression sustained between a first and second application of the same load, with an intervening reduction to the initial pressure.

This comparatively straight part of the figure is more or less inclined towards the axis of abscissas; the greater the angle, or the closer the straight part approaches the axis of ordinates, the greater is the rigidity or stiffness of the specimen. The approximate straightness of the line shows that equal increments of load produce nearly equal amounts of compression, which proves that the material possesses elasticity, although only in an imperfect degree, since nearly every release of pressure shows some additional set. The piece does not recover its primitive length when first released from its load, and this shortening, or set, increases as the process of loading and releasing is carried on.

Owing to the brittleness, or rather deficiency in toughness, of freestone, it is difficult to tell precisely at what stage of the process the elastic limit is passed. It is here understood that *elastic limit* means that stress at which the compression ceases to be substantially proportional to the applied load, and increases at a greater ratio. It has sometimes been defined to be that point at which the first permanent set takes place, meaning the extension or compression, as the case may be, which remains after the stress that caused the lengthening or shortening of the piece has been removed.   Stoney says: "The limit of elasticity may be defined to be the greatest strain that does not produce a permanent set." Hodgkinson and Clark have found permanent set from very small loads; and this fact was corroborated by the experiments at Watertown.   It is true that false permanent set occurs with some material, meaning a permanent set that seems to be caused by a load within the elastic limit, but which disappears upon leaving the specimen unloaded for a short time, when the piece returns to its original length; this generally happens only with material more perfectly elastic than that under discussion.   A slight indication of false permanent set was observed, in the case of a 16-inch concrete cube of superior strength.   In "Notes on Building Construction," published by Rivingtons, London, Oxford, and Cambridge, it is said : "When such loads"—within the elastic limit—"are constantly repeated, though they may produce an inappreciable set as regards the original length of the bar, yet it is not an increasing set, does not lead to rupture, and may therefore practically be ignored. When, however, the load is greater than the limit of elasticity, an increasing set takes place upon each application, which eventually leads to rupture."

These views are quite pertinent to the subject under consideration. With rigid and imperfectly elastic material like freestone, useful aid for determining the elastic limit is furnished by comparing the successive increments of set during the progress of operations. After passing the primary set, which is always relatively considerable, the gradually increased load alternating with releases produces small but nearly equal increments of set as long as the total compression proceeds at a tolerably uniform rate. This fact is rather conspicuous in the larger cubes, where, due to the prolonged resistance, a considerable number of sets could be observed. During a certain period of straining and releasing, the sets continue at a comparatively regular rate; then a set of greater magnitude ensues, indicating that the limit of elasticity is passed. Professor Weyrauch, in "Strength and Determination of Dimensions of Structures of Iron and Steel," says: "The experiments of Bauschinger upon tension, compression, flexure, and torsion in every case indicated very precisely the elastic limit; for example, for tension, where for the same increment of load all at once a disproportionate extension occurred, the maximum of which was only obtained after some time. This sudden expansion is to be attributed almost entirely to permanent change of form (set); the transitory or non-permanent changes remain proportional to the stress until very nearly the limit of rupture, and the coefficient of elasticity is found to be always almost entirely independent of the stress."

The three largest sets of cubes were used to determine the modulus of elasticity of Haverstraw freestone, but in consequence of the difficulty of deciding upon the probable elastic limit, the results are simply approximate. The accompanying Table G gives the successive increments of compression and set of the several 10-inch, 11-inch, and 12-inch cubes, condensed from Special Table II. These data, in conjunction with the strain-diagrams, serve as the basis of an estimate of the modulus of elasticity.

TABLE G.

SHOWING GRADUAL COMPRESSION AND SET OF TEN-INCH, ELEVEN-INCH, AND TWELVE-INCH FREESTONE CUBES.

Size and Mark of Cube	Compression at 100,000 lbs	Additional Compression, from—						
		100,000 to 200,000 lbs.	200,000 to 300,000 lbs.	300,000 to 400,000 lbs.	400,000 to 500,000 lbs.	500,000 to 600,000 lbs.	600,000 to 700,000 lbs.	700,000 to 800,000 lbs.
10-inch, a..	.0220″	.0170″	.0120″	.0090″	.0085″	....	....	....
10-inch, b..	.0145″	.0085″	.0075″	.0075″	.0105″	.0083″	....	....
10-inch, c..	.0132″	.0088″	.0080″	.0090″	.0085″	....	....	....
10-inch, d..	.0157″	.0093″	.0075″	.0087″	.0108″	....	....	....
Mean ..:	.0163″	.0109″	.0087″	.0085″	.0096″	....	....	....
11-inch, a..	.0152″	.0108″	.0080″	.0072″	.0073″	.0077″	....	....
11-inch, b..	.0145″	.0095″	.0064″	.0072″	.0070″	.0080″	....	....
11-inch, c..	.0170″	.0100″	.0080″	.0070″	.0080″	.0075″	....	....
11-inch, d..	.0140″	.0088″	.0072″	.0088″	.0112″	.0100″	....	....
Mean ...	.0152″	.0098″	.0074″	.0075″	.0084″	.0083″	....	....
12-inch, a..	.0185″	.0097″	.0073″	.0075″	.0067″	.0068″	.0065″	.0070″
12-inch, b..	.0130″	.0075″	.0060″	.0053″	.0050″	.0060″	.0070″	.0085″
12-inch, c..	.0192″	.0096″	.0067″	.0065″	.0065″	.0075″	.0098″	....
12-inch, d..	.0110″	.0075″	.0063″	.0052″	.0055″	.0065″	.0075″	.0070″
Mean ...	.0154″	.0086″	.0066″	.0062″	.0059″	.0067″	.0077″	.0075″

Size and Mark of Cube	Set at 100,000 lbs.	Additional Set, from—							Total Crushing Strength. Pounds.
		100,000 to 200,000 lbs.	200,000 to 300,000 lbs.	300,000 to 400,000 lbs.	400,000 to 500,000 lbs.	500,000 to 600,000 lbs.	600,000 to 700,000 lbs.	700,000 to 800,000 lbs.	
10-inch, a..	.0130″	.0100″	.0045″	.0025″	....	....	....	....	520,000
10-inch, b..	.0062″	.0018″	.0020″	.0017″	.0025″	....	....	....	650,500
10-inch, c..	.0049″	.0022″	.0019″	.0025″	.0022″	....	....	....	800,000 +
10-inch, d..	.0032″	.0026″	.0021″	.0033″	.0048″	....	....	....	644,000
Mean ...	.0073″	.0041″	.0026″	.0025″	.0032″	....	....	....	....
11-inch, a..	.0075″	.0045″	.0032″	.0027″	.0018″	.0027″	....	....	791,000
11-inch, b..	.0060″	.0022″	.0023″	.0015″	.0020″	.0015″	....	....	785,000
11-inch, c..	.0080″	.0038″	.0022″	.0020″	.0018″	.0032″	....	....	779,200
11-inch, d..	.0052″	.0026″	.0021″	.0033″	.0048″	.0040″	....	....	769,000
Mean ...	.0067″	.0033″	.0024″	.0024″	.0026″	.0029″	....	....	....
12-inch, a..	.0085″	.0030″	.0020″	.0015″	.0020″	.0018″	.0014″	.0023″	800,000 +
12-inch, b..	.0050″	.0020″	.0012″	.0016″	.0012″	.0018″	.0022″	.0030″	800,000 +
12-inch, c..	.0090″	.0030″	.0022″	.0018″	.0020″	.0020″	.0025″	....	764,000
12-inch, d..	.0035″	.0017″	.0013″	.0013″	.0007″	.0013″	.0017″	.0025″	800,000 +
Mean ...	.0065″	.0024″	.0017″	.0015″	.0015″	.0017″	.0019″	.0026″	....

The weakest of the 10-inch cubes (*a*) shows from the beginning much greater compression and set than any of the other pieces. Considerable internal strain, causing rapid change of form, is revealed by the amount of permanent set as loading progresses: the set is about three times greater than for the other samples; the total compression also is much more considerable. The strongest cube (*c*), which did not fail under the maximum load of 800,000 pounds, exhibited quite a uniform rate of compression from 100,000 to 500,000 pounds, when the micrometer was taken away: it probably maintained a similar rate to a much greater pressure; it evidently possessed in a remarkable degree the quality of being "homogeneous as to strain," as termed by Professor Thurston.

The other two cubes, *b* and *d*, which were of medium strength, kept rather close together as regards rate of shrinkage under pressure, up to about 400,000 pounds; within that range they suffered about equal amounts of compression and set.

Cubes *a* and *c* represent, therefore, the minimum and maximum strength of the 10-inch freestone cubes; *b* and *d*, which are of medium strength, are well suited to decide, approximately, where the elastic limit may be located. Their successive increments of compression from 200,000 to 400,000 pounds do not vary sensibly from a uniform rate; but each shrinks more rapidly between the latter load and 500,000 pounds. The same relation is observed with the permanent sets. Examining also the average amounts of compression and set of the four cubes, an evident increase of both is found from 400,000 to 500,000 pounds; and we conclude that the limit of elasticity is probably at 400,000 pounds, or at a pressure of 4000 pounds per square inch, with an aggregate compression of 0".0494.

The four 11-inch cubes do not differ much from each other in ultimate strength, which varies from 760,000 pounds (cube *d*) to 791,000 pounds (cube *a*). They keep fairly abreast of each other in the progress of compression and set; at 600,000 pounds the weakest cube had shrunk 0.06 inch, or 12 per cent more than cube *b*, which had suffered the least amount of compression under that load. An inspection of the averages shows compression to progress about equally from 200,000 to 400,000

4

pounds; thence up to 600,000 pounds it also progresses regularly, but at a somewhat increased rate. The micrometer observations were not carried beyond the last-named load.

Elasticity.—The elastic limit of these cubes cannot be stated with any great degree of confidence.

For the four 12-inch cubes, also, the average gradual compressions furnish no distinct indication of the elastic limit, but there is an increase of set from 600,000 to 700,000 pounds, and still more so from 700,000 to 800,000 pounds. The limit may therefore be placed at 600,000 pounds, or at a load of 4166 pounds per square inch. The average total compression corresponding to that load is 0.0494 inch.

For computing the compressive modulus of elasticity of freestone, the data furnished by the 10-inch and 12-inch cubes are used.

The loads, within the elastic limit per square inch of bed, were 4000 pounds for a 10-inch cube, and 4166 pounds for a 12-inch cube, with aggregate amounts of compression of 0.0444 inch and 0.0494 inch, respectively.

Let $L =$ original length of cube in inches;

$l =$ compression within the elastic limit, by a force,

$f,$ in pounds per square inch of bed of cube;

and $E =$ modulus of elasticity of compression:

then $$l : L :: f : E,$$

$$E = \frac{L}{l} \times f.$$

We therefore have:

Modulus of elasticity for 10-inch cubes,   900,900 pounds.
Modulus of elasticity for 12-inch cubes,   1,012,000   "
Average modulus of elasticity of compression,   956,450   "

With a modulus of 956,450 pounds the elastic limit of 11-inch cubes would be near 500,000 pounds.

As a general result of these investigations, it may be stated that the elastic limit of freestone cubes averages about 65 per cent of their ultimate resistance. According to Weyrauch, K. Styffe found that with the most different varieties of iron and steel the ratio of elastic limit to ultimate strength lies ordinarily between $\frac{1}{1.4}$ and $\frac{1}{1.8}$, and even under the most unfavorable circumstances rarely falls below $\frac{1}{2}$.

Little information on the modulus of elasticity of stones is found in works on the strength of materials. In Stoney's "Theory of Strains" the modulus of white marble is given at 2,520,000 pounds (by Tredgold); of Holyhead quartz-rock on bed, 4,598,000; on edge, 545,000 (by Mallet); and that of Portland stone, a freestone of the oolitic variety of limestone, at 1,533,000 (by Tredgold).

After passing the elastic limit, equal additions of load produce constantly increasing amounts of compression and set, and with certain materials the curve becomes more or less concave towards the axis of abscissas. This terminal part of the diagram is well defined in mortars, concretes, and brickwork, where it gradually becomes approximately parallel to the base-line as the point of fracture is approached. With neat cement it is not so well developed; and with freestone it is almost imperceptible, except in a few instances.

The increasing rate of compression, after passing the elastic limit, is perhaps due to a loss of cohesion among the particles of the outer shell of the cube, especially of that part about midway between the two bed-faces, which yields by bulging or buckling on the line of least resistance; the available area of resistance in the cross-section, under continued and accumulating pressure, becomes, therefore, more and more reduced until fracture ensues.

The upper portions of the compression-diagrams of freestone cubes are generally rather straight, or are formed of an irregular broken line not greatly differing from a straight line, with the final part in several instances exhibiting a steeper ascent than the preceding portion. In some few cases a tendency to the formation of a final curve, concave toward the axis of abscissas, is traceable, as may be seen in the diagrams of 8-inch cubes *c* and *d*, and 9-inch cube *d*. The first-named piece broke under a load of 388,000 pounds, and the micrometer observations were carried up to that point. From 280,000 pounds to 360,000 pounds the diagram is almost a straight line; it then declines at 370,000 pounds, whence it slightly rises to 380,000 pounds, to incline again towards the axis of abscissas as fracture is approached. A similar formation of the terminal part of the diagram is noticed in 8-inch cube *d*; the final bending of the curve toward the base-line would probably have been still more marked if observations, which ceased at 387,000 pounds, had been continued to 395,700 pounds, the ultimate load. In the case of 9-inch cube *d*, micrometer observations were continued to the moment of fracture, which occurred under a pressure of 445,000 pounds. Here the terminal part of the diagram is convex toward the axis of abscissas from 400,000 to 420,000 pounds; the curve is then reversed, and becomes concave up to the breaking-point.

Three of the 12-inch cubes (*a*, *b*, and *d*) resisted the maximum pressure of 800,000 pounds, once applied. Their diagrams are practically straight lines up to that point, while cube *c*, which yielded under a load of 764,000 pounds, began to develop a slightly concave curve at 600,000 pounds, increasing its inclination toward the axis of abscissas from 700,000 to 740,000 pounds, when the last observation was taken.

The shortness of the concave bends where they exist, and their nearly complete absence in most other samples of freestone, indicates the rigidity and brittle character of that material, and the advisability, in building, of imposing upon it but moderate loads. During the process of loading there are scarcely any audible or visible indications of the effects of pressure, except what may be inferred from the readings of the micrometer. In every instance the piece failed suddenly; therefore the micrometer was removed as a matter of precaution at a comparatively early stage, except in a few cases, in which the fracture took place sooner than was anticipated.

According to the rules given by Professor Thurston (Report of the United States Board on testing iron, steel, and other metals), "a perfectly straight line beneath the elastic limit, perfectly parallel with the elastic line, shows the material to be

homogeneous as to strain, i.e., to be free from internal strains
such as are produced (in metals) by irregular or rapid cooling,
or by working too cold. Any variation from this line indicates
the existence and measures the amount of strain. A line considerably curved exhibits the existence of such strain."

With woods which Professor Thurston tested in regard to
their resistance to torsion, the autographic line of the diagram,
up to the elastic limit, is almost perfectly straight. With freestone, and to a less degree with mortars and concretes, the portion of the diagram referred to, and more especially its initial
part, shows by its convexity and by other irregularities the
defects of the material as regards homogeneity as to strain.

It is further stated as a rule, that "a line rising from the
elastic limit regularly and smoothly, approximately parabolic
in form, and concave toward the base-line, indicates homogeneity in structure, and the absence of such imperfections as
are produced in wrought-iron by cinder, or in cast metals which
have been worked from ingots, by porosity of the ingots. A
line turning the corner sharply when passing the elastic limit,
and then running nearly or quite horizontal, as in irons usually,
and in low steel, or actually becoming convex toward the base-
line, as with some of the woods, and then after a time resuming
upward movement by taking its proper parabolic path, indicates
a decided want of this kind of homogeneity."

The few instances in which the freestone diagrams beyond
the apparent elastic limit show a terminal curve which is more
or less concave toward the axis of abscissas sufficiently prove
that the material is deficient in homogeneity of structure. The
terminal curve of 9-inch cube *d* is at first convex, and then
bends over toward the axis of abscissas. Such irregularities, in
a less marked degree, are seen in 8-inch cube *c*. Indeed, varying capacity of resistance beyond the limit of elasticity, alternately diminishing and increasing, are indicated by the irregular
form of the upper part of the diagram of nearly every freestone
cube.

Resilience.—The strain-diagrams of freestone and other
material also serve to estimate their resilience, or the capacity
to resist suddenly applied loads or blows.

The resilience is measured by the continued product of a selected maximum resistance—either the crushing load, or the pressure at the elastic limit, or at some other point—by the corresponding amount of compression, and this by some co-efficient which varies from $\frac{1}{2}$ to $\frac{2}{3}$, according to the degree of toughness or ductility of the material. With strain-diagrams, resilience is represented by the area included between the curve, the ordinate of maximum pressure, and the axis of abscissas from the origin of the curve to the foot of the ordinate.

When a specimen is tested by gradually but continuously increasing the load until fracture takes place, the strain-diagram will be a continuous line from beginning to end. To compute the total resilience, the length of the axis of abscissas from the foot of the curve to the foot of the ordinate of ultimate pressure, the number of pounds of the latter and the value of the fractional coefficient are required.

Owing to the convexity of the initial or lowest part of the freestone diagram, some slight modification in the method of computation was thought justifiable. The extent of the area representing the resilience was considered to depend upon and to be restricted by the permanent set produced after applying a load about sufficient to relieve the material of that internal strain which is manifested by the aforesaid convexity, and by incidental irregularities seen in the lower portion of the diagram. That load may be regarded as of a preliminary character, causing the material to adjust itself for sustaining additional stress by rendering it more homogeneous as to strain, as far as its peculiar structure may permit. This preliminary pressure may be regarded as about equivalent in its effect to the practice of preparing railway girders for actual use by stretching under a heavy load, as mentioned by Stoney; to the relieving of metals from internal strain by annealing, heating, etc.; or in the case of very ductile metals, according to Professor Thurston, by "straining them while cold to the elastic limit and thus dragging all their particles into extreme tension, from which, when released from strain, they may all spring back into their natural and unstrained position of equilibrium."

The preparatory load required for freestone is much below

the elastic limit. It is simply the stress, after the application
of which the initial convex curve begins to merge into a comparatively straight line. In conformity with the preceding remarks, we may assume that by the gradual application of pressure those particles or groups of particles, under more or less
excessive tension or internal strain of some kind, are in a great
measure relieved from the same; and on removing the stress
and returning to the clamping load, i.e., the pressure necessary
to hold the piece securely suspended in the testing-machine,
those particles may be considered to be in a much more unstrained and natural relation with respect to each other. In
resuming the operation of loading we deal in fact with a somewhat modified specimen, the original length of which has been
slightly diminished by the amount of permanent set caused by
the preliminary stress.

To compute the resilience of a specimen, we have to examine the strain-diagram and determine the point where it begins
to assume the form of a straight line, or nearly so. For freestone cubes, 8 inches on a side and upwards, an initial load of
100,000 pounds was taken as the average pressure necessary to
bring the unbalanced particles of the stone into proper adjustment. The area representing the resilience is therefore considered to begin at a point on the axis of abscissas distant by
the length of permanent set produced by 100,000 pounds from
the foot of the curve. This method may be illustrated by referring more especially to 8-inch cube *c*, one of the two freestone
cubes the progress of compression in which was observed to the
final moment of fracture.

For this cube, Strain-sheet I. and Special Table I. show that
upon the second application of the load of 100,000 pounds, at
which moment the total reduction of its original length amounted
to 0.017 inch, with a permanent set equal to 0.0065 inch, the
convex curve begins to change to an approximately straight
line. The area of resilience is therefore measured from the
point on the axis of abscissas at a distance of 0.0065 inch from
the foot of the convex curve. The first part of this area is a
right-angled triangle, the altitude of which is the ordinate representing 100,000 pounds, and its base that portion of the

axis of abscissas extending from the foot of said ordinate to the point of first permanent set, equal in this case to $0''.0105$ $= (0''.017 - 0''.0065)$. The remainder of the area consists of trapezoids, the widths of which are the successive amounts of compression, and the heights the means of each successive pair of ordinates. The compression being given in parts of an inch, and the pressure in pounds, the resilience is expressed in inch-pounds.

An examination of the freestone diagrams shows that they generally become somewhat steeper as the cubes tested increase in size. Under equal loads, an 8-inch cube suffers more compression than a 10-inch or 12-inch cube, as may be expected from the fact that under such circumstances each unit of the smaller cube is subject to a greater strain than a unit of the larger one. In other words, under equal loads the larger cubes undergo less change of form and exhibit more stiffness. The following table (H) affords a comparison of the amount of resilience, under gradually increasing loads, of freestone cubes from 8 to 12 inches on a side, and of a pier composed of three 12-inch cubes with dry joints.

Some discrepancies will be noticed in the following table which are evidently due to variations in structure and strength of individual specimens, but on the whole the principle that the stiffness of the cubes increases with their size is sufficiently borne out.

TABLE H.

RESILIENCE OF HAVERSTRAW FREESTONE.

SIZE OF CUBES.

Amount of Loads in Pounds.	8″ × 8″ × 8″				9″ × 9″ × 9″				10″ × 10″ × 10″				11″ × 11″ × 11″				12″ × 12″ × 12″				Pier of 3 12″ Cubes a,b,d
	a	b	c	d	a	b	c	d	a	b	c	d	a	b	c	d	a	b	c	d	
100,000	660	520	525	510		460	440	530		430	440	490	395	450	473	440	535	400	550	385	1,985
150,000	1,285	1,214	1,150	1,214		1,061	1,077	1,405		942	950	1,052	1,057	1,012	1,069	990	1,210	869	1,100	841	3,335
200,000	2,266	2,135	2,125	2,128		2,026	2,090	2,730		1,700	1,745	1,840	2,025	1,840	1,940	1,800	2,005	1,625	1,910	1,540	5,225
250,000	3,452	3,331	3,250	3,405		3,069	3,139	4,102		2,521	2,589	2,684	2,902	2,382	2,817	2,587	2,759	2,244	2,641	2,215	7,104
300,000	5,509	5,137	4,760	5,013		4,572	4,595	6,020		3,675	3,770	3,835	4,275	3,640	3,890	3,850	3,839	3,150	3,535	3,100	9,400
350,000		7,483	6,632	7,182		6,037	6,060	7,645		4,845	5,151	5,184	5,282	4,775	5,127	5,117	4,805	3,962	4,391	3,912	11,626
400,000						8,312	8,266	10,098		6,550	6,735	6,980	6,645	5,990	6,540	6,740	6,130	4,900	5,810	5,050	14,195
450,000						11,052	9,731			8,232	8,541	9,249	8,090	7,477	8,134	9,035	7,447	5,962	7,191	6,325	17,021
500,000							11,506			10,891	10,520	11,570	10,055	9,390	10,260	12,000	9,170	7,150	8,985	8,000	20,430
550,000							12,691			13,256			11,892	11,621	12,045	14,575	9,824	8,775	10,822	9,525	23,842
600,000										16,861			14,685	14,666	14,300	17,950	12,935	11,050	13,435	11,600	28,180
650,000																	14,810	12,925	16,183	14,100	32,867
700,000																	17,675	15,650	20,345	17,150	40,030
750,000																	19,775	18,369		19,506	
800,000																	22,025	21,775		22,025	

Another Table (I) is submitted, which embodies the average results obtained with freestone, under gradually increased loads, with regard to its resilience per cube, per square inch of bed-surface, and per cubic inch of entire mass.

This table shows rather more strikingly the increased stiffness of cubes as they increase in size. . It also shows to what extent cubes are deficient in elasticity, and under which loads their behavior approaches to some extent the condition of perfect elasticity. A body, perfectly elastic, with a certain area of resilience under a given load, should develop four times that area when the load is doubled, since the compression would have progressed uniformly, and the areas are therefore proportional to the squares of the loads. We find, for instance, in the columns of resilience per square inch, that for an 8-inch cube under 100,000 pounds pressure the average resilience is 8.59 inch-pounds. If the stone were perfectly elastic, its resilience at 200,000 pounds should be 34.36 (8.59 × 4) inch-pounds; at 300,000 pounds it should be 77.31 (8.59 × 9) inch-pounds; and at 350,000 pounds, 105.37 (8.59 × 12.25) inch-pounds. The table gives, at the loads named, 33.56, 79.24, and 109.80 inch-pounds, respectively.

Adopting the resilience of freestone cubes at a pressure of 100,000 pounds as a basis for comparison, Table H shows that the resilience actually developed at the progressive stages of loading is generally below that due to a perfectly elastic condition, especially towards the closing part of the operation in each case, and with the larger cubes; another proof of the want of homogeneity of structure in this material.

In a number of cases it was not practicable to define the elastic limit, and consequently the resilience at that point. The total resilience at the crushing moment could, as already stated, be determined only for two of the freestone cubes. In several instances, the measurements for resilience were only carried up to a pressure considerably below the ultimate load.

Information of some importance in this matter is embodied in Table J. In introducing this table it must be remarked that the elastic limits given therein are merely approximations, and the

TABLE I.

AVERAGE RESILIENCE OF CUBES OF HAVERSTRAW FREESTONE PER CUBE, PER SQUARE INCH OF BED-SURFACE, AND PER CUBIC INCH OF SPECIMEN.

RESILIENCE IN INCH-POUNDS.

LOAD IN POUNDS	Of Cube, with Sides of—					Of Pier of 3 12" Cubes	Per Square Inch of Bed-surface of Cubes with Sides of—					Of Pier of 3 12" Cubes	Per Cubic Inch of Mass, in Cubes with Sides of—					Of Pier of 3 12" Cubes	LOAD IN POUNDS
	8"	9"	10"	11"	12"		8"	9"	10"	11"	12"		8"	9"	10"	11"	12"		
100,000	550	477	457	438	468	1,991	8.59	5.89	4.57	3.62	3.25	13.82	1.074	0.654	0.457	0.329	0.271	0.384	100,000
150,000	1,206	1,182	992	1,027	959	3,349	18.87	14.59	9.92	8.49	6.66	23.26	2.359	1.621	0.992	0.772	0.555	0.646	150,000
200,000	2,148	2,276	1,776	1,894	1,775	5,246	33.56	28.10	17.76	15.65	12.34	36.43	4.195	3.122	1.776	1.473	1.028	1.012	200,000
250,000	3,337	3,423	2,619	2,710	2,483	7,128	52.14	42.26	26.19	22.40	17.24	49.50	6.518	4.698	2.619	2.036	1.437	1.375	250,000
300,000	5,071	5,064	3,790	3,891	3,416	9,435	79.24	69.52	37.90	32.15	23.72	65.52	9.905	6.947	3.790	2.923	1.977	1.820	300,000
350,000	7,027	6,565	5,101	5,015	4,332	11,669	109.86	81.27	51.01	41.45	30.08	81.04	13.725	9.030	5.101	3.768	2.597	2.251	350,000
400,000		8,959	6,809	6,453	5,492	14,244		110.61	68.09	53.33	38.14	98.93		12.290	6.809	4.843	3.178	2.748	400,000
450,000		10,352	8,744	8,151	6,755	17,081		127.80	87.44	67.36	46.91	118.62		14.200	8.744	6.124	3.909	3.205	450,000
500,000		11,417	11,099	10,408	8,335	20,303		140.95	110.99	86.02	58.02	142.38		15.661	11.099	7.820	4.831	3.955	500,000
550,000		12,593	13,513	12,482	10,022	23,929		155.47	135.13	103.16	66.60	166.18		17.274	13.513	9.378	5.800	4.616	550,000
600,000			17,183	15,337	12,298	28,284			171.88	126.73	85.40	196.42			17.188	11.523	7.117	5.456	600,000
650,000					14,557	32,985					101.09	229.07					8.424	6.363	650,000
700,000					17,787	40,176					133.38	279.00					10.282	7.750	700,000
750,000					19,266						133.93						11.161		750,000
800,000					21,854						151.76						12.647		800,000

figures in the column of total or absolute resilience are, except in two cases previously mentioned, derived from calculations deduced from the area of resilience found at the pressure when the last micrometer observation was made, by assuming that this area in the case of a rigid body like freestone increases approximately with the square of the load.

TABLE J.

RESILIENCE OF CUBES OF HAVERSTRAW FREESTONE AT ELASTIC LIMIT, LAST MICROMETER OBSERVATION, AND AT CRUSHING LOAD.

SIZE AND MARK OF CUBE.	WITHIN ELASTIC LIMIT.			AT LAST OBSERVATION.		AT CRUSHING LOAD.		
	Load, Pounds.	Inch-pounds.		Load, Pounds.	Inch-Pounds.	Load, pounds.	Inch-pounds.	
		Individual.	Mean.				Individual.	Mean.
8″ — a....	240,000	3,146		330,000	7,175	397,000	10,309	
8″ — b....	280,000	4,325	3,381	370,000	8,631	438,400	11,096	10,518
8″ — c....	260,000	3,505		388,000	9,516	388,000	9,516	
8″ — d....	220,000	2,548		387,000	9,698	395,700	10,150	
9″ — b....	400,000	7,920		536,000	13,103	368,000	14,835	
9″ — c....	?	?	?	550,000	12,691	643,000	17,344	14,872
9″ — d....	?	?		445,000	12,438	445,000	12,438	
10″ — b....	440,000	7,810		640,000	19,335	630,500	20,396	
10″ — c....	?	?	7,395	500,000	10,570	800,000+	?	19,801
10″ — d....	400,000	6,980		550,000	11,570	644,000	19,206	
11″ — a....	500,000	10,055		600,000	14,685	791,000	25,538	
11″ — b...	500,000	9,390		600,000	14,665	785,000	25,094	
11″ — c....	600,000	14,300	10,121	600,000	14,300	779,200	24,111	26,053
11″ — d....	400,000	6,740		600,000	17,950	769,000	29,468	
12″ — a....	?	?		800,000	22,025	800,000+	?	
12″ — b....	?	?	?	800,000	21,275	800,000+	?	?
12″ — c....	600,000	13,435		740,000	24,089	764,000	25,671	
12″ — d....	?	?		800,000	22,025	800,000+	?	
3-12″ cubes	?	?		700,000	40,030	748,000	45,705	—

Sufficient power was lacking to crush three of the 12-inch cubes and one of the 10-inch cubes.

9-inch cube *a* and 10-inch cube *a*, the diagrams of which are very irregular, are omitted from the table.

Table J indicates that the capacity to resist blows safely, augments with the size of cubes. The mean resilience of four 8-inch cubes within the elastic limit—sometimes termed the proof-resilience—was found to be 3381 inch-pounds. The elastic resilience of 9-inch cubes was ascertained for only one specimen, for which it amounted to 7920 inch-pounds. This was, however, a rather strong sample of its kind.

The mean elastic resilience of two 10-inch cubes is 7395 inch-pounds. The mean proof-resilience of the four 11-inch cubes is 10,121 inch-pounds.

In Table K, expressing the absolute resilience in inch-pounds of freestone cubes of various sizes, the first line gives the number of inch-pounds, taken from Table J, the second the number that would result if the resilience were in proportion to the area of bed-surface ; and the third the number that would result if the resilience were proportional to the mass, taking for the second and third cases the average absolute resilience of an 8-inch cube as a basis for comparison.

TABLE K.

COMPARATIVE TABLE OF ABSOLUTE RESILIENCE OF FREESTONE CUBES.

	8″ Cube.	9″ Cube.	10″ Cube.	11″ Cube.
1. Resilience, as given in Table J	10,518	14,872	23,146	26,053
2. Resilience, if proportional to area of bed-surface	10,518	13,312	16,434	19,886
3. Resilience, if proportional to mass of cube	10,518	14,976	20,543	27,342

It will be seen from this table that the inch-pounds in the first and third lines agree so nearly as to suggest that *the absolute resilience of cubes of freestone and of kindred material may be approximately proportional to the mass of the cubes.*

The pier composed of three 12-inch freestone cubes, *a, b,* and *d*, further illustrates this matter. This pier was crushed under a load of 748,000 pounds ; the last micrometer observation was taken at 700,000 pounds, at which pressure the resilience was 40,030 pounds. Each of the three cubes had previously been subjected to the maximum stress of 800,000 pounds without fracture.

The effect of this preliminary compression is well illustrated by the diagram of the pier on Strain-sheet VIII. Its initial or lower part is but slightly concave, showing that whatever internal strain had existed in the cubes had been nearly removed by previous loading ; it then rises regularly with a gentle curve to near the point of fracture. The resilience developed by these cubes, singly as well as combined, at various stages of pressure from 100,000 pounds upwards, is found in Table H. It is seen that up to 200,000 pounds the area of resilience of the pier more or less exceeds the combined area of individual cubes *a, b,* and *d*; beyond this point the aggregate resilience of the three single cubes gradually grows larger than that of their combination. Under a stress of 700,000 pounds, this aggregate resilience amounts to 50,475 (17,675 + 15,650 + 17,150) inch-pounds, against 40,030 inch-pounds of the pier. The resilience of the pier of three cubes is therefore about $2\frac{1}{2}$ times as great as that of a single cube of the same kind. It seems reasonable to suppose that if no preliminary load had been applied to the cubes, and they had been well joined with a cementing substance,—in other words, if the pier had been a true monolith,—it would have shown a resilience equal to the aggregate resilience of three individual cubes, although its ultimate resistance to a dead crushing load would fall short of that of a single cube. 10,445 inch-pounds (50,475 — 40,030) expresses most probably the absolute loss of working strength of the cubes resulting from their having been already strained beyond their elastic limit, and from the absence of a binding or cementing substance in the joints.

These few observations would seem to indicate that *when the area of impact is equal to the area of bed-surface, the resilience of hard and rigid material like stone, when in the shape of prisms of the same form and area of cross-section but of varying heights, becomes greater as the height of the prisms increases, probably within limits depending on liability to flexure.* On the other hand, *the capacity to resist dead loads decreases with increasing height of the specimen, but increases when the height or thickness is reduced, this increase being especially rapid when the height of the prism is less than one half that of a cube of the same cross-section.*

The results would be entirely different if but a portion of the bed of the specimen were struck. A valuable series of experiments might be made to determine the comparative live and dead loads needed to fracture exactly similar specimens of stone, and also to show the effect produced by exposing only a part of the bed to the blow.

CHAPTER V.

TESTS OF CEMENTS.

THE phenomena attending the fracture of specimens of neat cement, and of mortars and concretes made with hydraulic cement, were in all essential features similar to those exhibited by Haverstraw freestone.

NEAT CEMENT.

The series of cubes and prisms of neat cement were formed of Dyckerhoff Portland cement. The specific gravity at the time of testing varied from 2.024 to 2.115, and averaged 2.068, assuming the weight of a cubic foot of water to be 62.5 pounds.* The weight of each piece is given in General Table II., but only the prisms and the cubes from 7 inches on a side upwards were actually weighed, the weight of the smaller cubes being computed. The age of the specimens when tested varied from 1 year 10 months and 3 days, to 1 year 11 months and 1 day; average, 1 year 10 months and 16 days.

The table gives the nominal and actual size of each piece, the method of testing (beds plastered or bare), age when crushed, compressive strength of specimen in pounds per square inch of bed-surface and per cubic inch of mass, with remarks relating to the behavior of the piece while being compressed. The actual dimensions of a cube or prism generally varied by small fractions of an inch from the nominal size. The computations of crushing load per square inch of bed-surface and per

---

* As the memorandum-book of the late Mr. Cocroft, who had charge of the manufacture of the specimens, could not be found, it is not known what the raw cement which was used for the cubes weighed. Subsequently, a cask of Dyckerhoff cement was weighed at Fort Tompkins. The gross weight was 394.75 pounds; the cement weighed 371 pounds, with a volume of 3.42 cubic feet. The weight of a struck bushel would therefore be 135 pounds.

cubic inch of mass are based upon the actual dimensions of
each piece.

Inspection of the fragments showed the mass to contain
numerous globular cavities like blow-holes or air-bubbles, from
the size of a small pin-head to those having a diameter of $\frac{1}{8}$, or
even $\frac{3}{16}$ of an inch. As these cubes and prisms were made
with great care, the formation of such cavities was probably
unavoidable.

The faces of the larger cement cubes, previous to being
tested, exhibited an infinite number of minute hair-cracks, visible only on moistening the piece. During the latter stages of
the tests, signs of approaching destruction were given by the
appearance of many irregular cracks upon the surface of the
exposed sides, followed by a scaling or blistering off of thin
sheets or slabs, occasionally of quite considerable area. These
phenomena are not only indicative of the outward pressure on
the line of least resistance, but also of the probability that the
outer skin of the artificial stone had during the process of setting acquired a higher degree of density, hardness, and rigidity
than the interior mass. The harder outer crust did not compress as readily and rapidly as the core, and therefore cracked,
and under the strain which it suffered from the bulging mass
of the interior was detached and forced away from the body
of the piece. The hair-cracks upon the surface more or less
facilitated the separation of scales.

Cubes of Neat Cement.—The cubes varied by increments
of an inch from one inch to twelve inches on a side. There
were six samples of each size. The six 1-inch cubes, one 2-inch
and one 3-inch cube, five of the 11-inch cubes and the six 12-
inch cubes, were tested with their bed-faces plastered. The
other cubes were crushed without plaster finish, because the
2-inch cubes were the first tested, the 1-inch cubes as well as
one 2-inch and one 3-inch cube having been temporarily set
aside on account of slight irregularities of form. When the
set of 11-inch cubes was reached, the discrepancies between the
amounts of pressure required to crush the individual samples
of the preceding sets gave rise to the suspicion that the bed-
faces of the specimens might not have a sufficiently uniform

bearing against the pressing head-plates of the testing-machine. To remedy this supposed evil the beds of the remaining cubes, and of all the cement prisms, were plastered.

The average crushing load per square inch of bed-surface was 5000 pounds.

From General Table II., which contains the essential details noted while testing Dyckerhoff cement, Table L is condensed. It gives the observed crushing loads per square inch

TABLE L.

COMPRESSIVE STRENGTH OF CUBES OF DYCKERHOFF CEMENT.

The cubes marked * had their beds plastered.

Side of Cube.	Mark.	Crushing Load per Square Inch. Of Specimen.	Average.	Excess or Deficiency of observed load in relation to 5,000 pounds.
1 inch	a	*5,657 pounds		
1 "	b	*5,931 "		
1 "	c	*5,902 "		
1 "	d	*5,652 "	5,896 pounds.	Excess, 15.2 per cent.
1 "	e	*6,059 "		
1 "	f	*6,176 "		
2 inches	a	8,221 pounds		
2 "	b	6,525 "		
2 "	c	6,130 "		
2 "	d	7,261 "	7,094 pounds.	Excess, 29.5 per cent.
2 "	e	6,307 "		
2 . "	f	*8,218 "		
3 inches	a	5,997 pounds		
3 "	b	5,578 "		
3 "	c	5,772 "		
3 "	d	5,634 "	5,937 pounds.	Excess, 15.8 per cent.
3 "	e	5,840 "		
3 "	f	*6,795 "		
4 inches	a	5,138 pounds		
4 "	b	5,395 "		
4 "	c	4,335 "		
4 "	d	5,481 "	4,847 pounds.	Deficiency, 3.15 per cent.
4 "	e	4,612 "		
4 "	f	4,123 "		
5 inches	a	4,145 pounds		
5 "	b	4,594 "		
5 "	c	4,927 "		
5 "	d	4,786 "	4,610 pounds.	Deficiency, 8.5 per cent.
5 "	e	5,040 "		
5 "	f	4,170 "		

TABLE L.—(*Continued.*)

COMPRESSIVE STRENGTH OF CUBES OF DYCKERHOFF CEMENT.

Side of Cube.	Mark.	Crushing Load per Square Inch.		Excess or Deficiency of observed load in relation to 5,000 pounds.
		Of Specimen.	Average.	
6 inches...... ...	a	3,972 pounds		
6 "	b	3,582 "		
6 "	c	4,401 "	4,283 pounds.	Deficiency, 16.7 per cent.
6 "	d	4,975 "		
6 "	e	3,762 "		
6 "	f	5,003 "		
7 inches..........	a	4,554 pounds		
7 "	b	3,849 "		
7 "	c	5,134 "	4,987 pounds.	Deficiency, 0.26 per cent.
7 "	d	5,774 "		
7 "	e	5,180 "		
7 "	f	5,429 "		
8 inches..........	a	4,488 pounds		
8 "	b	4,629 "		
8 "	c	4,540 "	5,007 pounds.	Excess, 0.14 per cent.
8 "	d	5,597 "		
8 "	e	5,533 "		
8 "	f	5,255 "		
9 inches.	a	4,574 pounds		
9 "	b	4,594 "		
9 "	c	4,889 "	4,754 pounds.	Deficiency, 5.18 per cent.
9 "	d	4,783 "		
9 "	e	5,736 "		
9 "	f	3,946 "		
10 inches..........	a	3,902 pounds		
10 "	b	5,859 "		
10 "	c	5,123 "	4,761 pounds.	Deficiency, 5.02 per cent.
10 "	d	4,225 "		
10 "	e	4,710 "		
10 "	f	4,747 "		
11 inches..........	a	4,820 pounds		
11 "	b	*5,208 "		
11 "	c	*5,895 "	5,374 pounds.	Excess, 6.96 per cent.
11 "	d	*5,451 "		
11 "	e	*5,585 "		
11 "	f	*5,287 "		
12 inches..........	a	*4,910 pounds		
12 "	b	*5,379 "		
12 "	c	?		
12 "	e	*5,532(?) "		
12 "	f	*5,343 "		

The nominal 12-inch cube *d* is omitted, because in moulding it an error occurred, causing its bed to measure 12″ × 11.″3, instead of 12″ × 12″. The cubes marked * had their beds plastered.

of bed-surface of the individual cubes: the average for the several sets; and the percentage of excess or deficiency of the latter when compared with the average crushing load of 5000 pounds per square inch of bed-surface.

The individual crushing loads of the 1-inch cubes vary but little from their average; the same is true of the five plastered 11-inch cubes, and probably also of the five 12-inch cubes if sufficient machine power had been available to break cubes c and e at the first application of pressure. This indicates the good effect of plastering the bed-faces.

The average resistance per square inch of bed-surface of the 1-inch cubes is nearly 1200 pounds less than that of the 2-inch cubes, while the average strength of the 3-inch cubes is 1157 pounds less, or about the same as that of the 1-inch cubes. The only plastered 3-inch cube (f) showed the greatest strength in its set, being about 14½ per cent stronger than the average, and about 10 per cent stronger than the strongest of the five un-plastered cubes of the same set.

From the 2-inch cubes to the 6-inch cubes the average strength per square inch decreases; it then rises in the 7-inch and 8-inch cubes, again decreases in the 9-inch and 10-inch sets, and increases for the 11-inch and 12-inch cubes, but without developing the resistance offered by the 1-inch cubes.

12-inch cube c was not immediately broken on reaching the ultimate available load of 800,000 pounds, although pieces began to fly off at 770,000 pounds; it rapidly failed, however, and was destroyed when the maximum load had been sustained for about thirty seconds. Two other cubes, c and d, of the 12-inch series, did not yield when the maximum load was first reached, although cracks became visible at about 700,000 pounds. In these cases fracture was caused by reducing the load to the initial pressure of 5000 pounds and then gradually raising the pressure to 800,000 pounds.

From Table L it is seen that the average crushing load of the five unplastered 2-inch cubes is 6869 pounds per square inch, while the one 2-inch cube (f) that had been plastered only failed under a load of 8218 pounds; the rates being as 100 to 119.6. Unplastered cube (a) showed, however, a strength of 8121 pounds.

The five unplastered 3-inch cubes developed an average strength of 5764 pounds per square inch of bed-surface, while one plastered cube (*f*) broke under a load of 6795 pounds; the ratio being 100 to 117.9.

The five plastered 11-inch cubes vary about 13 per cent from one another in strength; their average is nearly 14 per cent greater than unplastered cube *a* of this set.

Finishing the beds of cement specimens with a thin layer of plaster seems to have brought out their strength as fully as any amount of machine-finishing would have done.

Prisms of Neat Cement.—The smallest of the cement prisms, $4'' \times 4'' \times 1''$, yielded under an average pressure of 261,104 pounds, equivalent to 16,320 pounds per square inch of bed-surface (Table E). When removed from the press, the sides of the prisms were found to have been forced out all round in the shape of irregular but approximately triangular bodies, leaving an apparently solid core formed of two short truncated pyramids, firmly adhering to each other. On removing the shattered lateral fragments the edges of the beds broke away, leaving the bases of the pyramids with less area than the original prisms.

Comparing the mean resistance per square inch of bed-surface of these $4'' \times 4'' \times 1''$ prisms with that of the 1-inch cement cubes, the average strength of which was 5896 pounds (Table L), the prisms are found to be 2.76 times as strong as the cubes.

This ratio is different when the $4'' \times 4'' \times 2''$ prisms are compared with the 2-inch cement cubes. The prisms yielded under an average aggregate load of 101,920 pounds, or 6370 pounds per square inch, while the 2-inch cubes show an average ultimate resistance of 7094 pounds per square inch. The cubes are therefore over 10 per cent stronger than the prisms. The exceptional strength of the 2-inch cement cubes has already been noted. It is not impossible that with a greater number of specimens of either form the ratio would have been different.

The average strength of the three $4'' \times 4'' \times 3''$ prisms is 6003 pounds per square inch of bed-surface, while that of the 4-inch cubes is only 4847 pounds. But the latter were crushed without plastered heads, while this preliminary treatment had been applied to the prisms. It has been shown that by plastering the beds the strength of the cubes is more fully brought out; and in order to make as fair a comparison as practicable, we therefore select the strongest of the unplastered 4-inch cubes *d*, which had a crushing resistance of 5481 pounds per square inch. On this basis the prisms show 9½ per cent. more strength than the cubes.

Table M exhibits in condensed form the strength of the different 4″ × 4″ prisms when compared with the strongest of the 4-inch cement cubes, the strength of the latter being taken as unity.

TABLE M.

Size of Prism.	Crushing Strength.	
	Per Square Inch.	Relative.
4″ × 4″ × 1″,............	16,320 pounds	2,978
4″ × 4″ × 2″,...........	6.370 "	1,162
4″ × 4″ × 3″,...........	6.003 "	1,095
4″ × 4″ × 4″ (strongest).	5,481 "	1,000

Table N gives a similar comparison of the strength of the 8″ × 8″ prisms with that of the strongest of the unplastered 8-inch cubes (*d*), the strength of the latter being taken as unity.

TABLE N.

Size of Prism.	Crushing Strength.	
	Per Square Inch.	Relative.
8″ × 8″ × 2″.....	10.664 pounds	1,923
8″ × 8″ × 3″.............	7.191 "	1,285
8″ × 8″ × 4″.............	5.952 "	1,064
8″ × 8″ × 5″.............	6,020 "	1,075
8″ × 8″ × 6″.....	5.771 "	1,031
8″ × 8″ × 8″.............	5.597 "	1,000

Both the 8-inch and 4-inch prisms show a striking increase of strength only when their height is reduced to one fourth of the cube of equal cross-section.

Four sets of prisms of neat cement, 12 inches square on bed, of heights of 2, 4, 6, and 8 inches respectively, had been prepared, there being three specimens of each set. The great resistance offered by some of the 12-inch cement cubes previously tested, rendered it improbable that any of these prisms could be crushed by the machine.

One of these large prisms of 2 inches thickness was tried and withstood the load of 800,000 pounds apparently without being affected by it in the least. The same occurred with one of the prisms 4 inches in thickness. It was then decided not to continue tests in that direction, but to ascertain the resistance of each set of three prisms formed as a dry-jointed pier.

The set of the three 12″ × 12″ × 2″ prisms resisted the maximum pressure of 800,000 pounds. The set of 12″ × 12″ × 4″ prisms (aggregating a little over 12 inches in height in the pier) failed under a load of 662,000 pounds. It is supposed that one of these prisms was in some manner defective, since the next larger pier of three 12″ × 12″ × 6″ prisms withstood a greater load. In this case the load was carried up to 700,000 pounds and then reduced to 5000 pounds. The driving-head of the machine was again put in motion, and the pier broke at 690,000 pounds, it evidently having been weakened by the first application of the pressure. The pier of 12″ × 12″ × 8″ prisms was crushed under a load of 654,800 pounds.

None of these last three piers showed as much resistance as the 12-inch cement cubes, while the 12″ × 12″ × 2″ pier of the same kind of material manifested superior strength, and only failed under a stress below the available maximum pressure when subsequently tested in conjunction with a 10-inch freestone cube.

COMPRESSION, SET, ELASTICITY, AND RESILIENCE OF DYCKER-
HOFF CEMENT.

[Special Table II., and Strain-sheets III. and IV.]

Compression and Set.—This cement is less subject to sudden fracture than freestone, and its general behavior during the last stages of the testing process, especially the unmistakable, visible and audible signs of impending disintegration, permitted a more prolonged use of the micrometer, which was in some instances kept on till fracture occurred.

The amount of set and compression generally with Portland cement is much less than with freestone. This cement is therefore decidedly stiffer than Haverstraw freestone. Under a load of 500,000 pounds the 11-inch freestone cubes show an average of over 71 per cent more compression than cement cubes of the same size; at 600,000 pounds, 57 per cent less. The 12-inch freestone cubes under pressures of 500,000, 600,000, and 700,000 pounds were compressed, in round numbers, 79, 63, and 46 per cent, respectively, more than similar cement cubes. Similar differences may be traced through the several sets of cubes of the two materials.

In the strain-curves of cement cubes the initial or lower parts are found to be much less convex toward the axis of abscissas than in the case of freestone. Especially is this true with the larger (11-inch and 12-inch) cubes.

The cement diagrams further disclose by their general form a more gradual yielding; the upper parts being better developed as regards concavity toward the axis of abscissas than in the case of freestone. In homogeneity as to strain as well as to structure, Dyckerhoff cement is superior to Haverstraw freestone, although inferior to it in absolute crushing strength. The irregularities of some of the cement diagrams, however, notably of 8-inch cube *d*, 9-inch cube *a*, 10-inch cube *d*, and 11-inch cubes *b* and *c*, prove that the material by no means possesses either kind of homogeneity in a superior degree.

Some of the cement diagrams are of especial interest.

8-inch cube *d*, broken by 360,000 pounds pressure, had the micrometer kept on until within 2000 pounds of that load, and therefore offers an opportunity to examine the strain-curve almost to the last moment. The irregularities of the upper or final branch of the curve, as it tends to take a direction nearly parallel to the axis of abscissas, exhibit both the destructive strain in progress and the deficiency of the piece in evenness of structure.

9-inch cube *c* gave decided indications of yielding after the load of 300,000 pounds had been exceeded. At 330,000 pounds one corner flew off ; at 350,000 pounds a crack appeared and the curve began to assume a direction approximately parallel to the axis of abscissas ; the cube did not yield, however, until a load of 396,000 pounds was reached.

9-inch cube *d* behaved differently.

The initial part of the diagram is nearly straight, from which it is concluded that the particles of the specimen were normally aggregated. Under higher pressure no indication was seen of approaching destruction ; some parts of the cube must have suddenly failed, and the ensuing jar probably caused a general giving way of the rest.

In the weakest of the 9-inch cubes (*f*) a lack of elasticity is noted, which is also indicated by the considerable amount of permanent set. The specimen failed under a pressure of 325,000 pounds, but began to crack at 130,000 pounds.

The strongest of the 10-inch cubes (*b*) broke under 587,200 pounds. It appears very rigid at the beginning, and somewhat abnormal in behavior. From 400,000 pounds up, however, the curve gradually bends downward, showing a proper successive yielding under increasing load.

10-inch cube *c*, the next strongest sample of its class, is quite different from the preceding piece. The diagram shows that it yielded rapidly at first, but that later on it displayed considerable stiffness.

The diagram of 10-inch cube *d* shows peculiar irregularities.

In some of the 11-inch cubes the initial part of the diagram is quite straight—a sign of homogeneity.

In 12-inch cube *a* set and elastic compression are regularly developed up to 500,000 pounds. The micrometer was kept on

TABLE O.

Cube.		Elastic Limit.			
				Average.	
Size.	Mark.	Load.	Compression.	Load.	Compression.
8-inch.......	b	220,000 lbs.	.0255″	212,000 lbs.	.0195″
8 "	c	200,000 "	.0182″		
8 "	d	260,000 "	.0203″		
8 "	e	200,000 "	.0180″		
8 "	f	180,000 "	.0153″		
9-inch.......	a	280,000 lbs.	.0244″	275,000 lbs.	.0215″
9 "	d	280,000 "	.0182″		
9 "	e	300,000 "	.0215″		
9 "	f	240,000 "	.0220″		
10-inch.	a	280,000 lbs.	.0221″	290,000 lbs.	.0192″
10 "	b	300,000 "	.0145″		
10 "	c	300,000 "	.0220″		
10 "	f	280,000 "	.0180″		
11-inch.......	a	280,000 lbs.	.0160″	421,666 lbs.	.0225″
11 "	b	400,000 "	.0122″		
11 "	c	450,000 "	.0255″		
11 "	d	500,000 "	.0250″		
11 "	e	500,000 "	.0250″		
11 "	f	400,000 "	.0212″		
12-inch.......	a	500,000 lbs.	.0248″	520,000 lbs.	.0265″
12 "	b	400,000 "	.0260″		
12 "	c	500,000 "	.0220″		
12 "	e	600,000 "	.0275″		
12 "	f	600,000 "	.0320″		

until the moment of fracture (710,000 pounds); and the diagram is interesting, as it fairly illustrates the gradual yielding of the material while approaching the ultimate load by bending down toward the axis of abscissas.

The 12-inch cubes c and e, and the nominal 12-inch cube d,* were not broken at the first application of the maximum load of 800,000 pounds, but only by repeating the process after returning to the initial load of 5000 pounds. Cube c exhibits irregular set up to 200,000 pounds, becoming more regular as the load increased to 400,000 pounds.

---

* This cube measured only 12″ × 11″.3 in cross-section, probably due to misplacement of one of the sides of the mould while inserting it.

TABLE

Resilience of Neat Dyckerhoff

Amount of Load in Pounds.	8" × 8" × 8"					9" × 9" × 9"						10" × 10" × 10"					
	b	c	d	e	f	a	b	c	d	e	f	a	b	c	d	e	f
100 000	460	450	410	425	440	515	650	525	305	380	450	490	210	515	500	400	350
150 000	1027	966	820	888	861	963	1172	1021	675	748	943	881	508	817	872	799	715
200 000	1973	1834	1438	1651	1545	1505	1754	1564	1185	1267	1696	1380	916	1218	1414	1286	1252
250 000	3108	2933	2257	2702	2293	2234	2409	2283	1917	1908	2545	1949	1372	1670	2035	1954	1814
300 000	....	....	3744	....	3833	3367	3216	3374	3127	2693	4770	2862	2087	2358	2869	3074	2589
350 000	....	....	5697	....	....	4798	4287	....	4204	3358	....	4306	2739	3170	...	4037	3659
400 000						....	....	....	....	....	4761	....	3755	4308	...	5960	4928
450 000						...	....	...	....	...	....	...	4862	5143	....	....	6484
500 000																	
550 000																	
600 000																	
650 000																	
700 000																	
750 000																	
800 000																	

The piece was crushed only under the sixth application of the maximum load of the testing-machine; the diagram presents practically a straight elastic line from 100,000 to 500,000 pounds.

Cube d (defective in size) required for its fracture four repetitions of the maximum load.

Cube e resisted a load of 800,000 pounds once, and showed great stiffness while it was again reloaded up to 800,000 pounds. It sustained that pressure for half a minute, and then yielded.

Elasticity.—Dyckerhoff Portland cement being stiffer than Haverstraw freestone, has a higher modulus of elasticity. Its elastic limit is somewhat less difficult to determine than that of freestone, although some cubes ran so irregularly as to render it unadvisable to consider them.

Table O gives the elastic limits of individual cubes, and the averages of sets of cubes.

P.

PORTLAND CEMENT.

SIZE OF CUBES.											PIER I. Of 3 Prisms each.	PIER II. Of 3 Prisms each.	PIER III. Of 3 Prisms each.
11″ × 11″ × 11″						12″ × 12″ × 12″					12″ × 12″ × 4″	12″ × 12″ × 6″	12″ × 12″ × 8″
a	b	c	d	e	f	a	b	c	e	f			
310	265	260	200	250	245	190	265	180	250	200	260	315	450
639	541	572	450	474	439	440	546	442	531	450	641	773	936
1092	979	1010	800	775	610	790	940	810	925	800	1175	1402	1701
1626	1479	1550	1306	1299	1190	1262	1435	1260	1375	1272	1946	2200	2678
2439	2107	2260	1925	1890	1850	1880	2040	1810	1925	1850	2840	3346	3822
3411	3027	3262	2730	2702	2662	2734	2706	2460	2656	2549	4010	4904	5445
4744	4099	4350	3550	3640	3600	3615	3475	3210	3450	3455	5260	7164	7320
5918	5436	5823	4862	4915	4912	4847	4750	4060	4512	4624	7280	10074	9844
8102	7285	7250	6145	6340	6290	6925	6175	5010	5700	5930	9180	13364	12705
.	9265	9057	8783	8724	8239	8515	8065	6322	7144	7552	11847	14045	16551
.	12084	11722	10905	11216	10690	10585	……	7760	8725	9340	15970	22246	21447
……	14868	14506	14660	……	……	14558	……	10084	11745	12166	.	27677	28576
……	18789	……	……	……	……	21013	……	12615	13983	16418	..	33938	……
								15876	16452	……	..		
										20647			

From the averages in Table O the average modulus of elasticity of Dyckerhoff Portland cement is found to be about 1,500,000, or, more correctly, 1,525,857.

Average modulus of elasticity for 8″ cubes, 1,358,774

"	"	"	"	" 9″ "	1,421,111
"	"	"	"	" 10″ "	1,510,416
"	"	"	"	" 11″ "	1,703,877
"	"	"	"	" 12″ "	1,635,107

Mean modulus of elasticity, ... 1,525,857

The modulus of elasticity of this kind of cement exceeds that of Haverstraw freestone by more than 50 per cent, and is practically identical with that of natural Portland stone (1,533,000), as determined by Tredgold.

Resilience.—Although the lower portions of cement strain-curves are less convex toward the axis of abscissas than

those of freestone, it is probably better to consider the process
of loading up to 100,000 pounds as merely preparatory, serving
to relieve the specimen of the greater part of existing internal
strain.   The area of resilience is therefore reckoned from that
point on the axis of abscissas representing the permanent set
when the first load of 100,000 pounds was reduced to 5000
pounds.

Table P exhibits the resilience in inch-pounds, under gradually increasing loads, of cement cubes from 8 to 12 inches on
a side, and of piers of cement prisms each 12 inches square on
bed, and of heights already described.

Owing to the imperfect elasticity of the material, no regular increase of the area of resilience proportional to the squares
of loads was found, but occasionally the actual development of
the area is nearly the same as it should be according to theory.
For instance, 8-inch cube *c* shows at 100,000 pounds a resilience of 450 inch-pounds; at 200,000 pounds, 1834, which by
theory should be 1800.   Eleven-inch cube *c*, with an area of
260 inch-pounds at 100,000 pounds, develops areas of 1010 and
2260 for loads of 200,000 and 300,000 pounds, respectively ;
theoretically, these areas should be 1040 and 2340.

Table Q gives interesting comparisons between the averages of the 12-inch cement cubes and the three piers of prisms.
The numbers of inch-pounds are given up to 600,000 pounds
for increments of 100,000 pounds.   For the two piers composed of 6-inch and 8-inch prisms respectively, two columns
appear in the table, one showing the observed resilience as developed under pressure, and the other the corresponding theoretical resilience, on the assumption that the resilience of specimens of the same cross-section but different heights varies as
the masses of the specimens.

It is seen that the shortest pier, composed of 3 prisms each
4″ in height, has larger areas of resilience than the corresponding average 12-inch solid cube; in fact, from 200,000
pounds upwards they are more than $1\frac{1}{2}$ times as large.   The
pier composed of the next larger prisms shows a similar excess
of actually observed resilience over that computed ; while with
the highest pier, representing in volume two 12-inch cubes

placed one on top of the other, the observed resilience is fairly comparable with that computed, except under the highest loads, and even then the difference is not considerable, and might have been less if an average could have been taken of several piers of that kind.

TABLE Q.

NEAT DYCKERHOFF PORTLAND CEMENT.

RESILIENCE OF 12" CUBES AND PIERS.

Pier I., composed of 3 prisms, each 12" × 12" × 4".
   " II.,   "   "   "   " 12" × 12" × 6".
   " III.,   "   "   "   " 12" × 12" × 8".

LOAD IN POUNDS.	12" Cubes. Observed Average.	RESILIENCE IN INCH-POUNDS OF—					
		Pier I.	Pier II.		Pier III.		
		Observed.	Observed.	Computed.	Observed.	Computed.
100,000	217	260	315	325	450	434
200,000	853	1,175	1,402	1,279	1,701	1,706
300,000	1,901	2,840	3,346	2,851	3,822	3,802
400,000	3,441	5,260	7,164	5,161	7,320	6,882
500,000	5,808	9,180	13,364	8,712	12,705	11,616
600,000	9,102	15,970	22,246	13,653	21,447	18,204

It is thought that the plastering of the bed-faces of all of these prisms had some influence on the results. Without going into details, it is obvious that a pier of three 12" × 12" × 4" prisms, coated in the aggregate with six thin layers of comparatively soft plaster, will compress more rapidly and show apparently more resilience than a solid 12-inch cube with two cushions only, or a pier of three 12" × 12" × 8" prisms. The latter had also six layers of plaster, but their aggregate thickness necessarily bore a lesser ratio to the collective height of the cement prisms than in the thinner pier, and compression proceeded therefore more slowly.

It was shown that at 700,000 pounds a dry pier of three 12-inch cubes of Haverstraw freestone exhibited about 2½ times the resilience of a single 12-inch cube, instead of three times; and the reason assigned for the difference was that the cubes had each previously been strained by a load of 800,000 pounds, which increased the stiffness, and that the pier was not a true monolith. The several cement prisms, with the exception of one measuring $12'' \times 12'' \times 6''$, had not previously been strained. Dyckerhoff cement is also less compressible than freestone; and the interposition of cushions of a more yielding substance, such as plaster of Paris from 36 to 48 hours old, will cause the combination of cement and plaster to develop more compressibility, and consequently more resilience, than without plaster.

It seems probable that with rigid material, divided into courses and subjected to compression, the interposition of a pliant and compressible binding substance essentially increases the capacity to resist concussions, or suddenly applied heavy loads.

With regard to the resilience of cubes of Haverstraw freestone, within the elastic limit, it was seen that there were indications that this property may increase with the size of the cube. This suggestion is to a certain extent corroborated by the results furnished by the cement cubes, as may be seen from Table R, which gives the resilience of the several cubes up to the elastic limit, the averages of the same for each class, both for the whole cube and per cubic inch of the mass.

A notable falling off in the amount of resilience is exhibited by the 10-inch cubes, which may perhaps be explained by the difficulty in many cases of determining the elastic limit. For this reason, several of the cubes are not recorded in the table. The 12-inch cubes evidently possess more of the property of resilience than the smaller ones, but their superiority in that respect is by no means marked.

It appears that for equal-sized cubes of Dyckerhoff cement and Haverstraw freestone, with equal striking weights, the safe height of fall is, for cements, on the average, a little more than one half that of freestone.

The fact that some of the cement cubes were plastered and some not, and that the micrometer was of necessity removed in most cases before the crushing load was reached, renders it unwise to try to deduce any conclusions as to the relative values of ultimate resilience of cement cubes of different sizes. With Haverstraw freestone cubes some evidence was shown that the ultimate resilience of cubes is proportional to their mass. The evidence with cement cubes is not sufficiently reliable to either prove or disprove this law.

TABLE R.

RESILIENCE OF CUBES OF DVCKERHOFF PORTLAND CEMENT WITHIN THE ELASTIC LIMIT.

Cube.		Resilience in Inch-pounds.			
Size.	Mark.	Load.	Of Cube.	Average.	Per Cube In.
8-inch.........	b	220,000 lbs.	2,288		
8 "	c	200,000 "	1,834		
8 "	d	260,000 "	2,423	1,876	3.66
8 "	e	200,000 "	1;651		
8 "	f	180,000 "	1,183		
9-inch.........	a	280,000 lbs.	2,753		
9 "	d	280,000 "	2,307		
9 "	e	300,000 "	2,783	2,539	3.48
9 "	f	240,000 "	2,312		
10-inch.........	a	280,000 lbs.	2,366		
10 "	b	300,000 "	2,087		
10 "	c	300,000 "	2,358	2,256	2.26
10 "	f	280,000 "	2,212		
11-inch.........	a	280,000 lbs.	2,004		
11 "	b	400,000 "	4,009		
11 "	c	450,000 "	7,250		
11 "	d	500,000 "	6,145	4,891	3.67
11 "	e	500,000 "	6,340		
11 "	f	400,000 "	3,600		
12-inch.........	a	500,000 lbs.	6,225		
12 "	b	400,000 "	3,475		
12 "	c	500,000 "	5,010	6,555	3.79
12 "	e	600,000 "	8,725		
12 "	f	600,000 "	9,340		

CHAPTER VI.

TESTS OF CEMENT MORTARS AND CONCRETES.

THE experiments made with these materials embraced tests of cubes of mortar and concrete of different sizes, and of different proportions of ingredients.

The following table gives the proportions of material that entered into the composition of the several mortars and concretes:

TABLE S.

COMPOSITION OF MORTARS AND CONCRETES.

Marks of Cubes.	Sizes of Cubes in Inches.	Kind of Cubes.	Kind of Cement.	Proportions by Volume.				Proportion of Cement o other Ingredients.
				Cement.	Sand.	Gravel.	Broken Stone.	
Fm	2, 4, 6, 8, 10, 12, 14, 16	Mortar	New'rk Co Rosendale Cement (dry measure)	1	3	..	..	1 to 3
Fc......	4, 6, 8, 10, 12, 14, 16, 18	Concrete	New'rk Co.'s Rosendale Cement (dry measure)	1	3	2	4	1 to 9
Am	4, 6, 8, 12, 16	Mortar	Norton's Cement (paste)	1	1½	..	..	1 to 1½
Ac......	4, 6, 8, 12, 16	Concrete	Norton's Cement (paste)	1	1½	..		1 to 7½
Bm	4, 6, 8, 12, 16	Mortar	Norton's Cement (paste)	1	3	..	..	1 to 3
Bc......	4, 6, 8, 12, 16	Concrete	Norton's Cement (paste)	1	3	..	6	1 to 9
Cm	4, 6, 8, 12, 16	Mortar	National Portland Cement (paste)	1	3	..	..	1 to 3
Cc......	4, 6, 8, 12, 16	Concrete	National Portland Cement (paste)	1	3	..	6	1 to 9

Two specimens of each kind and size of cubes had been prepared.

The age of the mortars and concretes marked F was about 22 months. The cubes of the combinations marked A, B, and C were older, and among themselves practically of equal age, varying only from 3 years 10 months and 4 days to 3 years 10 months and 14 days.

The beds of all of the cubes in Table S were plastered before being tested.

MORTARS AND CONCRETES OF NEWARK COMPANY'S ROSENDALE CEMENT.

In testing the mortar cubes of this cement, wooden pine cushions were placed between the plastered beds and the machine-heads, although former experiments indicated that the full strength of the material might not be brought out by this arrangement. The comparative roughness of the surfaces of mortar and concrete seemed, however, to call for the interposition of some comparatively soft material to secure a better equalization or distribution of the load over the pressed surface.

The thickness of the cushion-plates varied from $\frac{1}{4}$ inch to 1 inch, according to the size of the mortar cubes, which varied by increments of 2 inches from 2 to 16 inches on a side. The plates were square, and the length of their sides exceeded that of the sides of the cubes by about twice the thickness of the plate. The average weight per cubic foot was about 116 pounds for the mortar and 132 pounds for the concrete.

The crushing resistance of the individual specimens of each pair or set of mortar cubes was nearly the same, with the exception of the 2-inch and 10-inch cubes, the first differing from one another in strength per square inch about 27 per cent; the second, about 22 per cent. For the other sets, the greatest difference was not quite 5 per cent.

This satisfactory result with mortars suggested a change in the method of testing the series of concrete cubes of Newark Co.'s Rosendale cement.

One sample of each set was crushed with pine cushions, and the other directly between the machine-heads, it being thought that by the latter method superior compressive strength would be shown.

An opportunity for measuring the gradual compression and resilience of the concrete was thus afforded. The sides of these cubes were 4″, 6″, 8″, 10″, 12″, 14″, 16″, and 18″, re-

6

spectively. The cubes of the smallest set were both tested between wooden plates to see whether concretes crushed in this manner would give as uniform results as mortars. One cube broke under a pressure of 1074 pounds per square inch, the other at 991 pounds—a difference of 7.7 per cent.

When testing the other sets of concrete cubes, those crushed directly between the machine-heads proved in every instance stronger than their mates, which were broken between wooden cushions. On the average they exceeded them in strength nearly 19 per cent.

In testing one of the 10-inch, 12-inch, 14-inch, and 16-inch mortar cubes, respectively, the cushions were so placed that the directions of the grains crossed each other; in the other cases they were parallel.

In several instances, cleavage occurred on lines parallel to the grain whether the latter, in the two opposite plates, ran parallel or crosswise with respect to each other. The indentation of the wood cushions varied considerably in depth and uniformity. The stronger concretes caused deeper impressions in the wood than the mortars, the greatest observed depth being over $\frac{3}{10}''$ (10-inch concrete cube a); the maximum impression by mortar cubes ($\frac{22}{100}''$) occurred with 10-inch cube b. The observed cleavage of the material parallel to the grain of the wood indicates that wood cushions exercise a weakening influence upon the strength of stone. The fibre being forced sideways under pressure, undoubtedly reacts on the particles of stone with which it is in close contact, and favors their tendency to move laterally, in the direction of least resistance.

COMPRESSION, SET, ELASTICITY AND RESILIENCE OF MOR-
TARS AND CONCRETES MADE WITH THE NEWARK COM-
PANY'S ROSENDALE CEMENT.

Compression and Set.—The relative crushing resistance of cubes of mortar and concrete prepared with the Newark Company's Rosendale cement is shown in Table T, which gives ultimate pressures per square inch on bed-surface. The data are based on the figures in General Table III.

TABLE T.

COMPRESSIVE STRENGTH PER SQUARE INCH OF BED-SURFACE OF CUBES OF MORTAR AND CONCRETE, PREPARED WITH NEWARK COMPANY'S ROSENDALE CEMENT.

Composition of mortar: 1 vol. cement (dry measure), 2 vols. sand.
Composition of concrete: 1 vol. cement (dry measure), 3 vols. sand, 2 vols. gravel, 4 vols. stone.

	MORTARS.				CONCRETES.	
MARKS AND SIZES OF CUBES.	Strength in pounds per square inch.		How crushed—with Wooden Plates or Directly.	MARKS AND SIZES OF CUBES.	Strength in lbs. per sq. in. of piece.	How crushed—with Wooden Plates or Directly.
	Of Piece	Average.				
Fm 2" a.	1,653	1,429	W. P., grain parallel.			
" 2" b.	1,203	1,429	" "			
" 4" a.	752	758	" "	Fc 4" a.	1,074	W. P., grain parallel.
" 4" b.	765	758	" "	" 4" b.	991	" "
" 6" a.	818	800	" "	" 6" a.	1,025	" "
" 6" b.	782	800	" "	" 6" b.	1,230	Directly.
" 8" a.	701	707	" "	" 8" a.	876	W. P., grain parallel
" 8" b.	713	707	" "	" 8" b.	1,194	Directly.
" 10" a.	828	945	" "	" 10" a.	1,151	W. P. grain parallel.
" 10" b.	1,063	945	W. P., grain crosswise.	" 10" b.	1,182	Directly.
" 12" a.	699	685	" grain parallel.	" 12" a.	831	W. P., grain parallel.
" 12" b.	671	685	" grain crosswise.	" 12" b.	1,113	Directly.
" 14" a.	697	715	" grain parallel.	" 14" a.	698	W. P., grain parallel.
" 14" b.	733	715	" grain crosswise.	" 14" b.	748	Directly.
" 16" a.	613	612	" grain parallel.	" 16" a.	674	W. P., grain parallel.
" 16" b.	611	612	" grain parallel.	" 16" b.	1,030	Directly.
				" 18" a.	830	W. P., grain parallel.
				" 18" b.	1,041	Directly.

Among the mortars of the foregoing table, the 2-inch cubes have by far the greatest strength per square inch, about twice as much as the 8-inch, 12-inch, 14-inch, and 16-inch cubes; the last-named size is the weakest in the series.

Of the concretes the table shows that the cubes crushed without interposition of wooden plates are invariably stronger than those crushed between cushions, the average ratio being as 1080 to 871—a difference of about 19 per cent. When the series of mortar cubes from 4 inches to 16 inches on a side are compared with the corresponding concrete cubes which had' been broken between wooden cushions like the mortars, the strength of the concretes is superior to that of the mortars by about 15 per cent.

The average strength per square inch of the F mortar cubes, excluding the 2-inch cubes as being exceptionally strong, is 746 pounds. If no cushions had been used it might possibly have been about 19 per cent greater (the increase of strength found with the F concretes under such circumstances), or 888 pounds. The average strength of these concretes, crushed without cushions, was 1080 pounds; they are therefore about 18 per cent stronger than the mortars.

The comparison may be tabulated as follows:

Compressive strength per square inch of bed-surface of—

F mortar cubes, crushed between wooden cushions,   746 pounds;

F mortar cubes, crushed without cushions (estimated)................................. 888   "

F concrete cubes, crushed between wooden cushions................................... 871   "

F concrete cubes, crushed without cushions..... 1080   "

The use of wooden cushions in testing the F mortars, and one half of the F concretes, prevented the measuring of the gradual reduction of the length of the cubes under progressive compression. The micrometer was applied only in testing those F concretes that were crushed without interposition of wooden plates; i.e., one of the 10-inch, 12-inch, 14-inch, 16-inch, and 18-inch concrete cubes, respectively.

The strain-curves of the F concretes, and of the mortars and concretes marked A and B, made with Norton's cement, represented on Strain-sheets V. and VI., are characterized by the direction and form of the curve after passing the point where the elastic limit is located. The upper part of the curve here forms a decided bend, becomes concave toward the axis of abscissas, and then with a long sweep runs nearly straight and approximately parallel to that line until fracture takes place. With the material just named, micrometer observations could in most cases be continued until the end, or close to it, as no violent separation of parts took place, and cracks, if appearing at all, did so only just previous to disintegration. The final part of the curves proves that with these mortars and concretes the rate of compression augments rapidly under slight increments of pressure near the end of the operation. In this respect the curves are materially different from those of the cubes of freestone and neat cement (Strain-sheets I. to IV.), which indicate a considerable amount of rigidity as the ultimate load is approached.

An examination of the strain-curves of the *F* concretes renders it again apparent that during the initial stages of loading the compression of the smaller cubes progresses faster than that of the larger cubes. The marked breaks in the initial part of the diagram, especially of the 14″, 16″, and 18″ cubes, show that internal strains of some kind existed in the mass, caused probably by irregular setting after moulding. The groups of particles under abnormal internal strain were weakened and more or less disintegrated when a moderate pressure was applied from the outside.

Elasticity.—The data in Special Table III. and Strain-sheet V. approximately fix the modulus of elasticity of *F* concretes; the 14-inch cube was ignored, its diagram being too irregular.

Making proper allowance for the actual area of bed-surface and for the length of the cubes, we have for the formula,

$$E = \frac{L}{l} \times f.$$

For 10-inch *F* concrete cube, $L = 10''.22$; $l = .011''$; $f = 591$ lbs. $\left(\dfrac{60,000}{101.5}\right)$

For 12-inch *F* concrete cube, $L = 12''.02$; $l = .016''$; $f = 619$ lbs. $\left(\dfrac{90,000}{145.2}\right)$

For 16-inch *F* concrete cube, $L = 16''.16$; $l = .0175''$; $f = 619$ lbs. $\left(\dfrac{160,000}{258.4}\right)$

For 18-inch *F* concrete cube, $L = 18''.19$; $l = .0240''$; $f = 749$ lbs. $\left(\dfrac{230,000}{307}\right)$

Therefore,

Modulus of elasticity for 10 inch cube *F*.......................... 549,093 lbs.
Modulus of elasticity for 12-inch cube *F*.......................... 465,024 lbs.
Modulus of elasticity for 16-inch cube *F*......... .... 571,600 lbs.
Modulus of elasticity for 18-inch cube *F*.......................... 567,397 lbs.
Average modulus of elasticity of Newark Co.'s Rosendale cement
    concretes, approximately.................................. 538,349 lbs.

As might be expected, concrete compresses more rapidly than freestone or neat Portland cement. Comparing the 10-inch and 12-inch cubes of these materials, we have :

MATERIAL.	COMPRESSION IN INCHES UNDER PRESSURE OF—			
	50,000 lbs.	100,000 lbs.	150,000 lbs.	200,000 lbs.
10-inch Freestone Cube... ..	0.0092	0.0167	0.0207	0.0277
10-inch Cement Cube........	0.0056	0.0098	0.0127	0.0138
10-inch *F* Concrete Cube....	0.0088	0.0320	Exhausted.	....
12-inch Freestone Cube.....	0.0090	0.0183	0.0199	0.0244
12-inch Cement Cube... ...	0.0034	0.0057	0 0074	0.0098
12-inch *F* Concrete Cube....	0.0082	0.0210	0.0360	Exhausted.

In every case the rate of compression is much more rapid with concrete than with cement. Under 50,000 pounds pressure the length of the concrete cube is reduced about as much as that of freestone, but under greater loads the latter material shows greater resistance to compression.

Resilience.—The total resilience of cubes of concrete made with Newark Company's Rosendale cement is about half that of corresponding cubes of neat Dyckerhoff Portland cement. Their resilience within the elastic limit is small in comparison with their total resilience ; the material differs in that respect from Dyckerhoff cement and freestone. This is shown in Table U, which gives the loads and resilience, both within the elastic limit and at the moment of fracture ; also similar data for those cubes of neat cement and freestone whose ultimate resilience was directly measured. With the 12-inch and 16-inch concrete cubes the ultimate resilience was not directly measured. The 12-inch cube broke under a load of 161,600 pounds ; the micrometer was removed at 160,000 pounds, when the resilience amounted to 11,862 inch-pounds. The 16-inch cube broke under a load of 268,000 pounds, but the micrometer was taken off when the pressure had reached 260,000 pounds with an accumulated resilience of 18,219 inch-pounds. These differences of pressure being quite small, the final amounts of resilience were estimated, assuming the curve of the strain-diagram beyond the last measurement to be a true parabola.

The computed total resilience of the 12-inch concrete cube is 12,221 inch-pounds; that of the 16-inch cube, 19,586 inch-pounds.

TABLE U.

ELASTIC AND ULTIMATE RESILIENCE OF CUBES OF CONCRETE MADE WITH NEWARK COMPANY'S ROSENDALE CEMENT, OF NEAT DYCKERHOFF PORT-LAND CEMENT, AND OF FREESTONE.

Size of Cubes.	Material.	Elastic Resilience.		Ultimate Resilience.		Ratio of Elastic Resilience to Ultimate Resilience.
		Load in Pounds.	Inch-pounds.	Load in Pounds.	Inch-pounds.	
10-inch.........	*F* concrete.........	60,000	311	120,000	9,663	1 to 31.0
12 "	"	90,000	754	161,600	12,221	1 to 16.2
16 "	"	160,000	1,586	268,000	19,586	1 to 12.4
18 "	"	230,000	2,860	331,000	23,811	1 to 8.3
9-inch, *d*......	Dyckerhoff Portland Cement	280,000	2,307	390,000	5,760	1 to 2.5
11 " *d*......	" " "	500,000	6,145	674,000	18,157	1 to 3.0
11 " *e*.....	" " "	500,000	6,340	690,200	19,123	1 to 3.0
11 " *f*.....	" " "	400,000	3,600	645,600	15,198	1 to 4.2
12 " *a*.....	" " "	500,000	6,225	710,000	24,185	1 to 3.9
8 " *c*......	Haverstraw Freestone......	260,000	3,505	388,000	9,516	1 to 2.7

This table shows that cubes of freestone or Portland cement will probably safely resist for an indefinite number of times blows of a certain energy which represents a much larger fraction of their ultimate resilience (varying between $\frac{1}{2.5}$ and $\frac{1}{4}$) than concrete cubes of Newark Co.'s Rosendale cement. It would also appear that with these concrete cubes the ratio of elastic to ultimate resilience becomes greater as the size of cube increases; it must, however, be remembered that only one cube of each size was available for tests of this kind.

MORTARS AND CONCRETES OF NORTON'S CEMENT.

As shown in a preceding table (S) of this report, there were two kinds of mortar and concrete made with this cement, differing from each other in the proportion of sand used in making the mortar. The kind marked *A* was richer in cement, the proportion being 1 volume of cement paste to $1\frac{1}{2}$

volumes of sand ; for *B* mortar the proportion was 1 volume of cement paste to 3 volumes of sand. Six volumes of broken stone were added for concrete.

The following are the average weights and specific gravities of this material :

	Specific Gravity.	Weight per cubic foot.
A mortar	1.916	119.75 pounds.
A concrete	2.283	142.68 "
B mortar	1.871	116.94 "
B concrete	2.217	138.56 "

The age of these mortars and concretes when tested was a few days over 3 years and 10 months; they were therefore more than twice as old as those made of Newark Co.'s Rosendale cement. They were broken without interposition of wooden cushions. The cubes tested measured 4, 6, 8, 12, and 16 inches on a side, respectively ; there were two cubes of each size in every set of mortars and concretes.

The tests show that—

1. Mortars are generally not as strong as concretes made with those mortars.

2. The sets of mortars and concretes richest in cement proved stronger than the others.

3. The smallest (4-inch) cubes in each of the four sets were decidedly the strongest of the lot.

4. There is no apparent law of increase or decrease of strength per square inch of bed-area, as the size of cubes increases.

The foregoing statements are based on Table V, opposite.

Comparing the richer mortars and concretes (*A*) of Table V with each other, the average strength of all of the cubes of each material is about the same, but the concretes are stronger than the mortars in the 4-inch, 12-inch, and 16-inch cubes. In Class *B*, with a smaller proportion of cement, the concretes are on the average about 16 per cent stronger than the mortars. The richer *A* mortars show an average of nearly 45 per cent more strength than the *B* mortars : the *A* concretes 34 per cent more strength than the *B* concretes.

TABLE V.

COMPRESSIVE STRENGTH OF CUBES OF MORTAR AND CONCRETE MADE WITH NORTON'S CEMENT.

Am MORTAR. Composition: 1 vol. Cement and 1½ vols. Sand.	STRENGTH IN POUNDS. Per square inch of bed.	Average.	*Ac* CONCRETE. Composition: 1 vol. Cement, 1½ vols. Sand, and 6 vols. Broken Stone.	STRENGTH IN POUNDS. Per square inch of bed.	Average.
4-inch Cube, a	2,032	} 2,042	4-inch Cube, a	2,320	} 2,322
4 " " b	2,053		4 " " b	2,323	
6 " " a	1,378	} 1,340	6 " " a	909	} 963
6 " " b	1,303		6 " " b	1,016	
8 " " a	1,640	} 1,746	8 " " a	1,352	} 1,434
8 " " b	1,852		8 " " b	1,516	
12 " " a	1,326	} 1,346	12 " " a	1,503	} 1,560
12 " " b	1,366		12 " " b	1,617	
16 " " a	1,254	} 1,247	16 " " a	1,466	} 1,447
16 " " b	1,240		16 " " b	1,429	
Bm MORTAR. Composition: 1 vol. Cement, and 3 vols. Sand.			*Bc* CONCRETE. Composition: 1 vol Cement, 3 vols. Sand, and 6 vols. Broken Stone.		
4-inch Cube, a	1,483	} 1,324	4-inch Cube, a	1,551	} 1,633
4 " " b	1,166		4 " " b	1,715	
6 " " a	780	} 750	6 " " a	1,009	} 1,000
6 " " b	721		6 " " b	991	
8 " " a	848	} 790	8 " " a	879	} 861
8 " " b	732		8 " " b	844	
12 " " a	679	} 688	12 " " a	744	} 765
12 " " b	696		12 " " b	756	
16 " " a	749	} 718	16 " " a	858	} 843
16 " " b	687		16 " " b	828	

COMPRESSION, SET, ELASTICITY, AND RESILIENCE OF MORTARS AND CONCRETES MADE WITH NORTON'S CEMENT.

[Special Tables IV., V., VI., and VII., and Strain-sheets V. and VI.]

Compression and Set.—As samples of this class failed without explosive disruptions of spawls from the surface, the micrometer was used in most instances until the end of the operation.

The diagrams resemble those obtained with the concrete cubes of Rosendale cement. The initial part of the strain curve again discloses defective homogeneity in regard to strain—more strikingly so in the larger cubes than in the smaller ones.   The fact that all of the cubes were practically of the same age when broken may account for this result ; the seasoning of the smaller cubes was perhaps further advanced.

In nearly all of these diagrams the curve is at first convex toward the axis of abscissas ; it then ascends for a short length about tangentially to the convex curve, then bends over, forming a concave curve, and thus continues in nearly a straight course to the end, diverging but slightly from a direction parallel to the axis of abscissas.

The diagram of 12-inch mortar cube *a*, Class *A*, presents an exceptional appearance, quite different from its mate, 12″ cube *b*, and from the other samples generally.   It is from the beginning distinguished by a very rapid rate of compression with corresponding large sets.   When the load had risen to 50,000 pounds, the permanent set was 0.052 inch, or about 17 times as much as that shown by the companion cube under the same circumstances.   On reaching 100,000 pounds the set had increased to 0.076 inch—about 13 times the amount of set of the other cube.   From this point forward the curve, which thus far had been rather convex toward the axis of abscissas, is reversed and becomes concave, gradually changing to a nearly straight line when approaching the point of fracture.   Despite the uncommon rate of compression and set of this specimen, its ultimate strength was only about 3 per cent less than that of the other cube of the same size and class.   A part of the general giving way of the piece under pressure, especially during the first half of the operation, may possibly be ascribed to the fact that the plaster which coated the bed-faces was slightly thicker than in other cases.   The plaster at the close of the operation was found to be somewhat soft and yielding.   It is believed, however, that there must have been some more important cause : the cement may have been in a somewhat softer condition than in the other mortar cubes.

Elasticity.—The elastic limit is more distinctly marked in the diagrams of the larger *A* and *B* cubes than in those of the smaller ones. It must be borne in mind that the elasticity of mortar and concrete is far from being perfect; the irregularities of the diagrams, the numerous deviations from a straight line below the limiting point, and the considerable amount of permanent set observable at an early stage of the operation of testing, show that the term *elasticity* can be used here only in a restricted sense. For the limit of such imperfect elasticity as is peculiar to the artificial compounds in question, that point is taken at which the line of the diagram decidedly changes its former direction, with a tendency to incline toward the axis of abscissas.

In the 8-inch mortar and concrete cubes this change of direction occurs so gradually that it is difficult or impossible to fix upon any point as the elastic limit. A glance at the diagrams shows that this point is much more easily recognized in the 16-inch cubes, or even in the 12-inch cubes. These two kinds of cubes were therefore selected for determining the modulus of elasticity. omitting two cubes of Class *A*, viz., 12-inch mortar cube *a* on account of its abnormal behavior, and 12-inch concrete cube *b*, for which the point corresponding to the elastic limit cannot be recognized.

In Table W the approximate moduli of elasticity are obtained.

In each class the modulus of compressive elasticity of the concretes is higher than that of the mortars; within the elastic limit the concretes are therefore stiffer. The mortars and concretes of Class *A*, which contain a larger proportion of cement, are within that limit more rigid, or less compressible than those of Class *B*.

Resilience.—The total resilience of cubes of classes *A* and *B* could be directly observed and computed from actual measurement in twenty cases out of twenty-four. In the remaining four cases the micrometer observations were continued to within from $1\frac{1}{4}$ to 4 per cent of the ultimate load. For these cubes the probable area of resilience from the last point of direct observation to the final moment was computed by the method already explained.

TABLE W.

	Values of—			Modulus of Elasticity.
.	L	l	f	
A Mortar:				
For 12-inch Cube, b......	12″.11	0.0150″	$\frac{110,000}{144.5} = 761$	614,381 pounds.
" 16 " " a......	16′.13	0.0192″	$\frac{200,000}{256.2} = 781$	656,121 "
" 16 " " b......	16″.17	0.0182″	$\frac{200,000}{258} = 736$	653,908 "
Average.......... ..				641,470 pounds.
A Concrete:				
For 12-inch Cube, a......	12″.12	0.0130″	$\frac{110,000}{145} = 759$	707,621 pounds.
" 16 " " a......	16″.20	0.0160″	$\frac{200,000}{258.7} = 773$	782,662 "
" 16 " " b......	15″.27	0.0180″	$\frac{220,000}{257.4} = 855$	772,825 "
Average............				754,369 pounds.
B Mortar:				
For 12-inch Cube, a......	12″.08	0.0083″	$\frac{60,000}{145} = 414$	602,545 pounds.
" 12 " " b......	12″14	0.0090″	$\frac{60,000}{146.2} = 410$	553,044 "
" 16 " " a......	16″.10	0.0140″	$\frac{120,000}{259.4} = 463$	532,450 "
" 16 " " b......	16″.09	0.0172″	$\frac{120,000}{257} = 467$	436,862 "
Average............				531,225 pounds.
B Concrete:				
For 12-inch Cube, a......	12″.17	0.0080″	$\frac{60,000}{145.44} = 413$	628,276 pounds.
" 12 " " b......	12″.14	0.0075″	$\frac{70,000}{145.3} = 482$	780,197 "
" 16 " " a......	16″.21	0.0200″	$\frac{160,000}{258.7} = 618$	500,889 "
" 16 " " b......	16″.24	0.0180″	$\frac{160,000}{259.5} = 620$	559,378 "
Average............				616,935 pounds.

Table X shows the resilience of the several kinds of cubes made with Norton's cement.

TABLE X.

RESILIENCE IN INCH-POUNDS OF CUBES OF MORTAR AND CONCRETE MADE WITH NORTON'S CEMENT.

Kind of Material, Size and Mark of Cubes.	Load when Micrometer was removed.	Resilience when Micrometer was removed.	Ultimate Load.	Ultimate Resilience.	Average Ultimate Resilience.
A Mortar: 1 vol. Cement Paste, 1½ vols. Sand.					
8-inch Cube, a	106,000 pounds	1,913	106,000 pounds	1,913	} 2,288
8 " " b	120,000 "	2,663	120,000 "	2,663	
12 " " a	192,000 "	10,844	192,000 "	10,844	} 8,883
12 " " b	190,000 "	6,173	197,400 "	*6,923	
16 " " a	321,200 "	13,820	321,200 "	13,820	} 12,593
16 " " b	320,000 "	11,366	320,000 "	11,366	
A Concrete: 1 vol. Cement Paste, 1½ vols. Sand, 6 vols. Broken Stone.					
8-inch Cube, a	87,600 pounds	4,962	87,600 pounds	4,962	} 6,102
8 " " b	97,900 "	7,242	97,900 "	7,242	
12 " " a	215,400 "	14,700	218,100 "	*15,260	} 17,918
12 " " b	228,300 "	19,381	232,900 "	*20,576	
16 " " a	379,200 "	61,523	379,200 "	61,523	} 48,092
16 " " b	368,000 "	34,660	368,000 "	34,660	
B Mortar: 1 vol. Cement Paste, 3 vols. Sand.					
8-inch Cube, a	54,250 pounds	1,026	54,250 pounds	1,026	} 1,125
8 " " b	47,250 "	1,225	47,250 "	1,225	
12 " " a	98,500 "	3,197	98,500 "	3,197	} 3,186
12 " " b	101,600 "	3,175	101,600 "	3,175	
16 " " a	194,200 "	8,233	194,200 "	8,233	} 7,588
16 " " b	176,750 "	6,943	176,750 "	6,943	
B Concrete: 1 vol. Cement Paste, 3 vols. Sand, 6 vols. Broken Stone.					
8-inch Cube, a	54,300 pounds	1,678	56,400 pounds	*1,880	} 1,846
8 " " b	55,000 "	1,812	55,000 "	1,812	
12 " " a	112,650 "	7,101	112,650 "	7,101	} 4,657
12 " " b	109,900 "	4,657	109,900 "	4,657	
16 " " a	222,100 "	13,234	222,100 "	13,234	} 14,104
16 " " b	215,000 "	14,974	215,000 "	14,974	

In the foregoing table the figures denoting ultimate resilience marked * are estimated for the final part, the micrometer observations having in these cases not been carried quite up to the breaking-point.

The table proves clearly the superior resilience of concretes over the mortars which form their matrix; also, that this capacity of resisting concussion, etc., is much increased in mortars and concretes by increasing the amount of cement entering into their composition.

In Table Y the first line of numbers of inch-pounds of resilience, for each set or class, are the averages taken from Table X. The second and third lines give the figures which would obtain if the resilience were exactly proportional to the mass, as suggested in former parts of this report. The figures of the second line are based on the observed average resilience of the 8-inch cubes, and in the third line on the observed average resilience of the 12-inch cubes. A fourth line is added, which gives the averages of the second and third lines.

TABLE Y.

RELATING TO THE QUESTION WHETHER THE RESILIENCE OF CERTAIN BUILD-ING MATERIAL IS PROPORTIONAL TO ITS MASS, APPLIED TO CUBES OF MORTAR AND CONCRETE MADE WITH NORTON'S CEMENT.

KIND OF MATERIAL, ETC.	RESILIENCE IN INCH-POUNDS.		
	8-inch Cube.	12-inch Cube.	16-inch Cube.
A Mortar (1 vol. Cement, 1½ vols. Sand):			
1. Resilience according to Table X	2,288	8,883	12,593
2. Resilience, if proportional to mass, 8″ cube as basis	2,288	7,722	18,304
3. Resilience, if proportional to mass, 12″ cube as basis	2,632	8,883	21,056
4. Resilience, means of 2 and 3	2,460	8,302	19,680
A Concrete (1 vol. Cement, 1½ vols. Sand, 6 vols. Broken Stone):			
1. Resilience according to Table X	6,102	17,918	48,092
2. Resilience, if proportional to mass, 8″ cube as basis	6,102	20,594	48,816
3. Resilience, if proportional to mass, 12″ cube as basis	5,309	17,918	42,472
4. Resilience, means of 2 and 3	5,705	19,256	45,644
B Mortar (1 vol. Cement, 3 vols. Sand):			
1. Resilience according to Table X	1,125	3,186	7,588
2. Resilience, if proportional to mass, 8″ cube as basis	1,152	3,880	9,216
3. Resilience, if proportional to mass, 12″ cube as basis	944	3,186	7,552
4. Resilience, means of 2 and 3	1,048	3,533	8,384
B Concrete (1 vol. Cement, 3 vols. Sand, 6 vols. Broken Stone):			
1. Resilience according to Table X	1,846	4,657	14,104
2. Resilience, if proportional to mass, 8″ cube as basis	1,864	4,660	14,912
3. Resilience, if proportional to mass, 12″ cube as basis	1,742	5,879	13,955
4. Resilience, means of 2 and 3	1,803	5,269	14,433

Material deviations from the supposed law are seen only in the 16-inch cubes of *A* mortar. They are partially explained, as far as the figures of the first line of that series are concerned, by the high average of observed resilience of the 12-inch cubes (due to the great amount of resilience developed by 12-inch cube *a*, the abnormal behavior of which has already been commented on).

In the other three series of Table Y, considering the fact that in each class only two specimens of the same size of cube were available, the computed figures approach those derived from direct observation sufficiently near to increase the possibility of the truth of the law that resilience of cubes is proportional to the mass.

The concretes are greatly superior in resilience to the mortars which enter into their composition. The *A* concretes possess on the average about three times as much resilience as the *A* mortars; the *B* concretes about twice as much as the *B* mortars.

The advantage of a liberal proportion of cement in the composition of mortars is also clearly demonstrated. The richer mortars (*A*) possess about twice the resilience of the *B* mortars; and the richer (*A*) concretes an average of about 3.3 times that of the *B* concretes.

MORTARS AND CONCRETES OF NATIONAL PORTLAND CEMENT.

This cement was used in preparing one set of mortar cubes and one set of concrete cubes. Each set embraced 4-inch, 6-inch, 8-inch, 12-inch, and 16-inch cubes, respectively, there being two cubes of each size.

The mortar consisted of 1 volume of cement paste and 3 volumes of sand. To this mixture were added 6 volumes of broken stone for the concrete.

Specific gravity of mortar = 1.92; weight per cubic foot = 119 pounds.
Specific gravity of concrete = 2.249, weight per cubic foot = 140 5 "

The age of these cubes when broken was about 3 years 10 months and 5 days: identical, within a few days, with the age of the mortars and cements made with Norton's cement. They were tested without interposing wooden cushions.

The mortars and concretes of this cement are marked *Cm* and *Cc*, respectively, in the tables accompanying this report.

Table Z gives the observed crushing loads of the cubes, and the resulting averages, per square inch of bed-surface.

TABLE Z.

COMPRESSIVE STRENGTH OF CUBES OF MORTAR AND CONCRETE MADE WITH NATIONAL PORTLAND CEMENT.

Cm MORTAR. Composition; 1 vol. Cement, 3 vols. Sand.	STRENGTH IN POUNDS.		*Cc Cement.* Composition: 1 vol. Cement, 3 vols.Sand, 6 vols. Broken Stone.	STRENGTH IN POUNDS.	
	Per square inch of bed.	Average.		Per square inch of bed.	Average.
4-inch Cube, *a*......	3,612	} 3,450	4-inch Cube, *a*......	3,923	} 4,014
4 " " *b*......	3,288		4 " " *b*.....	4,105	
6 " " *a*......	2,768	} 2,655	6 " " *a*......	2,436	} 2,629
6 " " *b*......	2,542		6 " " *b*......	2,823	
8 " " *a*......	2,586	} 2,469	8 " " *a*......	3,058	} 3,025
8 " " *b*.....	2,353		8 " " *b*......	2,993	
12 " " *a*......	2,472	} 2,434	12 " " *a*......	2,540	} 2,690
12 " " *b*......	2,396		12 " " *b*......	2,840	
16 " " *a*......	2,501	} 2,519	16 " " *a*......	2,880	} 2,978
16 " " *b*......	2,537		16 " " *b*.. ...	3,077	

The concretes carry a heavier dead load than corresponding mortars by about 13.5 per cent. The smallest cubes are again the strongest, relatively, in their set; the 4-inch mortar cubes exceed by 27 per cent the average strength per square inch of the other cubes of their set; the 4-inch concrete cubes exceed the average of the other concrete cubes by 29 per cent.

An opportunity is here afforded to note the influence of the quality of the cement upon the compressive strength of mortars and concretes. Class B of mortars and concretes prepared with Norton's cement is in every respect, including age, identical with Class C, for which National Portland cement was used. Comparing the average crushing loads per square inch of bed-surface of the latter class of samples (Table Z) with those of Class B (Table V), we find that the National Portland cement mortars are fully three times as strong as the Norton mortars, and the same ratio exists between the concretes. The *C* mortars and concretes are also stronger than those of the *A* class of Norton's cement, although the latter contain twice as much cement. The *C* mortars exceed the average strength of the *A* mortars by 75 per cent; the *C* concretes surpass the *A* concretes fully 100 per cent.

As Norton's cement enjoys a good reputation in the market, these results speak well for the brand known as National Portland cement.

COMPRESSION, SET, ELASTICITY, AND RESILIENCE OF MORTAR AND CONCRETE MADE WITH NATIONAL PORTLAND CEMENT.

[Special Tables VIII. and IX., and Strain-sheets VI. and VII.]

Compression and Set.—The rate of compression was measured for the 8-inch, 12-inch, and 16-inch cubes, both mortars and concretes. In every instance the micrometer observations were continued to the moment of fracture. The superior compressive strength and stiffness of National Portland cement mortars and concretes, compared with the corresponding cubes of the two classes of mortars and concretes of Norton's cement, are quite apparent when the strain-sheets are inspected. The National cement shrinks less under equal loads than the cubes of the Norton cement classes, and after passing the elastic limit, which, however, can be but roughly located, the final sweep of the strain-curve to the terminal point is much shorter and more curved than with the *A* and *B* specimens.

The existence of internal, unbalanced strain, successively overcome in the first stages of loading, is indicated in the *C* mortars by the irregular broken line presented by the diagrams in rising up from the axis of abscissas. There are slight traces of convexity toward the axis of abscissas, excepting with 8-inch cube *b*. Deficiencies in homogeneity of structure, nearly up to the point of fracture, are especially noted in 12-inch mortar cube *a*.

7

The C concrete cubes are also defective in homogeneity, both as to strain and as to structure, but in a lesser degree than the mortars. The final sweep of the strain-curves toward the breaking-point is comparatively much longer than with the mortars—an indication of greater tenacity. 12-inch cube a, Strain-sheet VII., is remarkable for the sudden change of direction of the line at 160,000 pounds; the elastic limit is here clearly defined.

Both in strength and in general configuration of diagrams, the National Portland cement mortars and concretes form a sort of medium between those made with Norton's cement on one side, and neat Dyckerhoff Portland cement on the other. The data of gradual compression contained in the Special Tables (Table II. for the neat cement, and Tables VIII. and IX. for the C cubes) from which the strain-diagrams were constructed show that in the 8-inch C cubes compression proceeds at about the same rate as in the 8-inch cement cubes up to 100,000 pounds; but when this load was reduced to 1000 pounds the permanent set of the C mortars averaged about $1\frac{1}{2}$ times that of the cements, that of the concretes $2\frac{1}{2}$ times. The C compounds suffer. therefore, more permanent change of form than the cements. Beyond 100,000 pounds the compression and set of the mortars, and still more that of the concretes, proceed at a faster rate than that of the cements.

For the 12-inch cubes, neat cement and C mortars compress at about equal rates up to 200,000 pounds; further on, the superior rigidity of the Dyckerhoff cement asserts itself. The 12-inch concretes compress throughout more rapidly than the 12-inch cement cubes. At 200,000 pounds their average permanent set is $0''.0075$ against $0''.0022$ for neat cement; at 300,000 pounds the average set of the concretes is $0''.0248$, or just eight times as much as that of the cements.

Elasticity.—It is with difficulty, and with considerable doubt as to the correctness of the results, that the modulus of elasticity of the C cubes is determined. From the Special Tables and Strain-sheets the following table is prepared :

TABLE A₁.

MODULI OF ELASTICITY OF CUBES OF MORTAR AND CONCRETE MADE WITH NATIONAL PORTLAND CEMENT.

$$E = \frac{L}{l} \times f.$$

KIND AND SIZE OF CUBES.	Breaking Load. Pounds.	Limit of Elasticity Pounds.	Area of Bed. Sq. ins.	L	l	f Pounds.	Modulus of Elasticity Pounds.
C Mortar:							
8-inch cube, a	168,000	130,000	64.96	8″.13	.0122″	2,001	1,333,453
8 " " b	150,000	110,000	63.76	8″.12	.0165″	1,725	862,500
12 " " a	357,000	240,000	144.60	12″.15	.0125″	1,729	1,680,590
12 " " b	345,600	240,000	144.24	12″.15	.0132″	1,664	1,531,636
16 " " a	650,000	460,000	259.85	16″.24	.0140″	1,770	2,053,500
16 " " b	654,500	480,000	257.90	16″.20	.0180″	1,861	1,674,900
Average							1,522,665
C Concrete:							
8-inch cube, a	196,500	110,000	64.24	8″.24	.0132″	1,712	1,068,700
8 " " b	193,500	70,000	64.64	8″.21	.0082″	1,083	1,084,200
12 " " a	367,000	160,000	144.48	12″.19	.0100″	1,107	1,349,433
12 " " b	410,000	240,000	144.36	12″.18	.0150″	1,663	1,350,360
16 " " a	747,000	440,000	259.40	16″.19	.0170″	1,695	1,614,238
16 " " b	800,000 +	480,000	260.00	16″.24	.0162″	1,846	1,850,558
Average							1,386,248

If we compare the averages of this table with the average modulus of elasticity of Dyckerhoff cement, we find that the C mortars are in that respect identical with the cement, while the modulus of the concretes is about 10 per cent lower.

With regard to Norton's cement mortars and concretes of Class B, which have in composition the same proportions as the C cubes, it is found that, within the elastic limit, the B mortars compress three times as much as the C mortars, and the B concretes about twice as much as the C concretes.

There is some doubt as to whether these average moduli express exactly the elastic status of the material. The last table shows a gradual rise of the modulus as the sizes of cubes increase. The same occurs, though in a much less marked degree, in the *A* mortars and concretes, but the reverse occurs in those of Class *B*. With the Dyckerhoff cement cubes, 8-inch, 9-inch, and 10-inch cubes have the lowest moduli, and the 11-inch and 12-inch cubes the highest. The modulus of the 10-inch freestone cubes is about 12 per cent lower than that of the 12-inch cubes.

The diagrams show distinctly that in every case the initial or lower part of the strain-curves of the lesser cubes is more inclined toward the axis of abscissas than that of the larger cubes, or that their rate of compression under equal loads is greater. When the limit of (imperfect) elasticity is reached with the larger cubes, their compression has not advanced as much in proportion to their size as that of the smaller specimens at the same point, and this circumstance may account for the difference in the moduli.

Resilience. —The micrometer having been kept on to the end of the operation for every piece prepared with National Portland cement, the ultimate resilience could be directly measured. Two of the twelve cases under consideration are rather exceptional. 8-inch mortar cube *b* broke when the load of 150,000 pounds had been put on a second time. From 100,000 to 150,000 pounds the set was 0″.0045, or as much as from 1000 pounds to 100,000 pounds. When the pressure of 150,000 pounds was reached the first time the micrometer showed a compression of 0.025 inch ; on the second application of the same load the compression increased to 0.031 inch, and the piece failed. The other case is 16-inch concrete cube *b*, which proved quite refractory.

When the available maximum load of 800,000 pounds had been put on there were no signs of impending fracture.

The piece was only broken upon a fifth application of the maximum load. The details connected with this experiment are discussed farther on.

The following Table B shows the approximate amounts of resilience of mortar and concrete cubes *C* at the elastic limit and at the crushing load :

TABLE B₁.

RESILIENCE AT ELASTIC LIMIT AND AT CRUSHING LOAD OF CUBES OF MORTAR AND CONCRETE MADE WITH NATIONAL PORTLAND CEMENT.

Composition: *Mortar C =* 1 vol. Cement Paste, 3 vols. Sand.
"   *Concrete C =* 1 vol. Cement Paste, 3 vols. Sand, 6 vols. broken Stone.

MATERIAL AND SIZE OF CUBES.	RESILIENCE OF ELASTIC LIMIT.			RESILIENCE AT CRUSHING LOAD.		
	Load. Pounds.	Inch-pounds.		Load. Pounds.	Inch-pounds.	
		Of Cube.	Average.		Of Cube.	Average.
C Mortar:						
8-inch cube, *a*	130,000	803	} 752	168,000	2,154	} 1,993
8 " " *b*	110,000	702		150,000	1,832	
12 " " *a*	240,000	1,422	} 1,512	357,400	5,957	} 6,207
12 " " *b*	240,000	1,603		345,600	6,437	
16 " " *a*	460,000	3,515	} 4,087	630,000	10,451	} 12,627
16 " " *b*	480,000	4,660		654,500	14,803	
C Concrete:						
8-inch cube, *a*	110,000	472	} 362	196,500	6,548	} 6,422
8 " " *b*	70,000	252		193,500	6,297	
12 " " *a*	160,000	644	} 1,095	367,000	18,505	} 16,293
12 " " *b*	240,000	1,546		410,000	14,082	
16 " " *a*	440,000	4,012	} 3,973	747,000	47,316	} 65,223
16 " " *b*	480,000	3,934		800,000+	83,130	

The absolute resilience of the concretes is again far superior to that of the corresponding mortars. The *C* mortars are about twice as resilient as the *B* mortars, which have the same proportion of sand ; and the *C* concretes are about four times as resilient as the *B* concretes. In absolute resilience, classes *A* and *C* are about equal ; *A* having twice the amount of cement (Norton's) in its composition that *C* has.

With respect to resilience at the elastic limit, the National Portland cement cubes are decidedly superior to those of Norton cement : but the *C* mortars possess somewhat more resilience than the *C* concretes, while with Norton cement the reverse is the case. It is possible that if more samples had been available these relations might have been changed.

Using the averages of total resilience, as given in Table B₁,

Table C, is formed, to investigate whether the *C* class of cubes conform to the problematic rule that the resilience of cubes is about proportional to their mass.

TABLE C_1.

RELATING TO THE QUESTION WHETHER THE RESILIENCE OF CERTAIN BUILD-
ING MATERIAL IS PROPORTIONAL TO ITS MASS.   APPLIED TO CUBES OF
MORTAR AND CONCRETE MADE WITH NATIONAL PORTLAND CEMENT.

KIND OF MATERIAL, ETC.	RESILIENCE IN INCH-POUNDS.		
	8-inch Cube.	12-inch Cube.	16-inch Cube.
C Mortar (1 vol. Cement. 3 vols. Sand) :			
1. Resilience according to Table C_1..........................	1,993	6,207	12,627
2. Resilience, if proportional to mass, 8″ cube as basis.....	1,993	6,726	15,944
3. Resilience, if proportional to mass, 12″ cube as basis....	1,839	6,207	14,713
4. Resilience, means of 2 and 3............................	1,916	6,466	15,328
C Concrete (1 vol. Cement, 3 vols. Sand, 6 vols. Broken Stone):			
1. Resilience according to Table C_1.................... ...	6,422	16,293	65,223
2. Resilience, if proportional to mass, 8″ cube as basis.....	6,422	21,674	51,376
3. Resilience, if proportional to mass, 12″ cube as basis....	4,828	16,293	38,620
4. Resilience, means of 2 and 3............................	5,625	18,983	44,998

There is a notable divergence in the 16-inch cubes, both in mortars and concretes. For the mortars, the highest calculated amount of resilience, line 2, exceeds the observed one by nearly 21 per cent; the lowest, line 3, by about 14 per cent. For the concretes, the highest calculated resilience, line 2, is about 21 per cent less than the observed one, while the lowest figure, line 3, falls short by 41 per cent. With the mortars, the discrepancies are not generally very great; with the con-cretes it should be noted that the high average of observed resiliences of 16-inch cubes is due to the extraordinary resist-ance of 16-inch cube *b*, which developed nearly twice as much resilience as 16-inch concrete cube *a*. If the computed amount of resilience of the 16-inch concrete cubes, lines 2, 3, and 4, are compared with the observed resilience of 16-inch cube *a* (47,316 inch-pounds: see Table B,), we find the agreement be-tween the several figures quite close.

The peculiar features of the breakage of 16-inch concrete cube *b* were as follows : The diagram plainly shows that when the maximum load had been reached the first time the elastic limit had already been passed. The total compression at that time was 0".053. Returning to the initial load of 5000 pounds, a permanent set of 0".027 was noted ; it had therefore recovered but one half of the loss of length caused by the first maximum load. Putting pressure on again, the compression was measured at intervals of 100,000 pounds. The lower part of this second diagram is slightly concave toward the axis of abscissas, showing some internal strain, still existing ; thence it rises in a nearly straight line of less inclination than presented by the first diagram up to 700,000 pounds ; the stiffness and elasticity had evidently increased. At 700,000 pounds the first crack appeared in sight, and up to 800,000 pounds the diagram bends downward, though but slightly. The power of resistance was evidently not exhausted ; this was also shown by the very moderate increase of compression (0".007) at 800,000 pounds, and of permanent set (0".005) on returning again to 5000 pounds.

During the third loading observations were made only at 400,000, 600,000, 700,000, and 800,000 pounds. The rather more pronounced concavity of the upper branch of the diagram shows that the cube had begun to yield, though slowly. The total compression when the maximum load was put on a third time was 0".0665 ; the piece was allowed to rest under that load for 10 minutes, at the end of which time the reduction of its length had progressed to 0".0752, an increase of 0".0087.

Reducing the load to 5000 pounds, the permanent set now amounted to 0".0415 ; it was visibly increasing. The piece was now allowed to rest under this minimum load for 6 minutes, during which time it actually recuperated slightly, recovering 0".001 of its length, the total set at the end of the period being 0".0405.

When loading was resumed, compression was again measured at every 100,000 pounds. The augmented inclination of the diagram toward the axis of abscissas generally, the increasing convexity of the lower part and more decided concavity of the upper, indicate approaching destruction. The cube was again left for 10 minutes exposed to the maximum stress of 800,000 pounds; the compression increased from 0".081 at the beginning to 0".093 at the end of that time.

When the pressure was reduced for the last time to 5000 pounds, the piece was left under this minimum stress for 6 minutes. At first the permanent set was 0".055; this, after 4 minutes, was reduced to 0".0532, which was still recorded at the 6th minute.

Pressure was once more put on, and measurements taken at every 100,000 pounds. Decided convexity at the lower end, a rather straight line for the middle portion, and concavity at the upper end characterize the last diagram. When 800,000 pounds was reached a fifth time a total compression of 0".102 was recorded. After remaining under the maximum pressure for 2 minutes the cube yielded quite rapidly and broke to pieces. The whole operation had lasted one hour and twenty minutes.

The question naturally arises what the ultimate load of this cube, once applied, might have been if the testing-machine had possessed sufficient power to determine it. It seems that an approximate estimate can be formed by knowing how much resilience was developed by the piece, and assuming that as much would have been shown by it if loading had steadily progressed up to the point of fracture. The terminal parts of the strain-diagrams of the other five concrete cubes made with National Portland cement are all similar to each other, and it is entirely probable that if sufficient power had been applied the diagram of 16-inch cube *b* would not have been materially different from the others, especially not from that of 16-inch cube *a*. A rough computation made with these premises shows that the actual crushing load would probably have been about 900,000 pounds, corresponding to a strain-curve which would represent about the same area of resilience as was developed by repeating the maximum load of the machine four times.

The series of operations necessary to break the concrete cube just described suggests another more important line of tests. Wöhler's experiments, made under the auspices of the Prussian Government in the years from 1858 to 1870, and then continued by Spangenberg, have shown that iron and steel can be ruptured under pressures considerably below their ordinary breaking loads, by repeating the pressure a sufficient number of times.

In calculating the dimensions of different parts of a structure the usual method is to adopt some factor of safety, so that each piece is strained only a fractional part of its ultimate strength. This fraction is made smaller for live loads than for steady stresses. Wöhler's experiments were designed to ascertain the maximum stress, with various amounts of minimum load, which could be repeated an indefinitely great number of times without injuring the piece. By using a fraction of this limit, a new and apparently more scientific and rational factor of safety would be obtained. The conclusion based upon the experiments referred to, known as Wöhler's laws, have since been formulated by Launhardt, Weyrauch, and others; also in Appleton's Cyclopædia of Applied Mechanics. It has been remarked, however, by authors writing on the subject, that Wöhler's experiments, although extensive, do not furnish decisive results. It is quite certain that the extension of researches of this kind to cements, mortars, concretes, etc., has not yet been thought of.

An obvious reason for the incomplete condition of these investigations is the tediousness of loading and unloading a single test-piece a great number of times, as was done by Wöhler. To use the testing-machine at the Watertown Arsenal for such purposes would be out of the question. A practical alternative would seem to consist in preparing a liberal number of samples of some material which should be divided into several sets. One set should be used to find the average ultimate strength, once applied, noting general behavior, limits of elasticity, resilience, and any other points of interest. The samples forming the second set should each be subjected to a stress a certain percentage less than the ultimate strength, recording the number of times such stress had to be repeated to produce fracture. The pieces of the other sets would be treated similarly, reducing for each consecutive set the terminal load in a certain ratio. By such a system of approximation it might be possible to determine both graphically and by formulæ the average compressive load which might be safely repeated a very great number of times; such tests would occupy but a moderate length of time.

CHAPTER VII.

TESTS OF BRICK PIERS.

THE sets of brick piers tested comprised six piers, all of the same size, 1½ brick in cross-section and six courses high. They were built up of common hard, North River brick, laid in hydraulic mortar made of 1 part of Newark Co.'s Rosendale cement, and 2 parts of sand. The mortar-joint averaged about ¾ of an inch thick. Each pier had a base and cap of North River bluestone, of the same cross-section as the pier, with their bed-faces rubbed smooth and plane. The height of the brick-work between the bluestone varied from 16 to 16½ inches; the length of the piers varied from 22 to 23¼ inches, including the end stones.

The age of the piers when broken was 1 year 9½ months.

The results of the tests are found in General Table VI. and in Compression or Special Table X.; they are graphically represented on Strain-sheet VIII.

The first indications of destructive strain were sharp, snapping sounds at a comparatively early part of the operations. Longitudinal cracks appeared later, at loads averaging about 80 per cent of the crushing load. The cracks would generally follow the line of joints, first on one side and then on the other. On approaching the ultimate load, cracks were also formed at other places. During the later stages of the operation an almost continuous grinding, crackling noise was heard, sounding as if fire was raging in the pier.

The diagrams of the brick piers resemble those of the mortars and concretes of the Norton cement classes, except that the curves of the brickwork are somewhat more regular. It is not thought that the interposition of the bluestone flags had an appreciable influence upon the form of the brick strain-curves, since bluestone is far superior in strength to brickwork, and would in the form of prisms of only a few inches in thickness experience but little change of form at the load which destroyed the pier. All of the bluestone flags were perfectly sound when the broken piers were removed from the machine.

The crushing strength of the piers varied from 250,000 to 291,000 pounds, and averaged 266,587 pounds, equivalent to 1851 pounds per square inch, or 119 gross tons per square foot. The following table gives a comparison of the breaking strength of the piers and the 12-inch cubes of the several mortars and concretes, tested without wooden cushions; the 12-inch cubes being selected as being nearest in size to the brick piers:

TABLE D_1.

COMPRESSIVE STRENGTH OF BRICK PIERS AND OF CUBES OF MORTAR AND CONCRETE.

Brickwork : 12″ × 12″ in cross-section, 6 courses high.
Cubes of mortar and concrete : 12 inches on a side.

NOTE.—C = Cement, S = Sand, Gr = Gravel, Bk = Broken Stone.

MATERIAL.	COMPOSITION.				Strength in lbs. per square inch.	
	C	S	Gr	Bk	Of Piece.	Compared with brick pier
Brick pier..........	......	.....	...	......	1,851	100
Concrete cube F...............................	1	3	2	4	1,113	60
(Made with Newark Co.'s Rosendale cement.)						
Mortar cube Am..............................	1	1½	....	....	1,346	72.7
Concrete cube Ac...............................	1	1½	....	6	1,560	84.3
Mortar cube Bm...............................	1	3	....	......	688	37.2
Concrete cube Bc.............................	1	3	....	6	765	41.3
(Made with Norton's cement.)						
Mortar cube Cm...............................	1	3	....	......	2,434	131.5
Concrete cube Cc..............................	1	3	....	6	2,690	145.3
(Made with National Portland cement.)						

The brick piers were stronger than concretes made with Newark Co.'s Rosendale cement, and the mortars and concretes made with Norton's cement, but weaker than those made with National Portland cement.

The micrometer was kept in use to the crushing-point, except for pier No. 1, from which it was removed at 280,000 pounds, while the pier broke at 291,000 pounds. Table E, gives the data of resilience at the elastic limit and at the crushing load.

TABLE E₁.

RESILIENCE OF BRICK PIERS.

Piers: 12 inches square, 6 courses (16″ to 16½″) high; bluestone cap and base. Common hard North River brick. Mortar: 1 vol. Newark Co.'s Rosendale Cement; 2 vols. Sand.

NUMBER OF PIER.	RESILIENCE AT ELASTIC LIMIT.			RESILIENCE AT CRUSHING LOAD.		
	Load, Pounds.	Com-pression.	Inch-pounds.	Load, Pounds.	Com-pression.	Inch-pounds.
No. 1.............	170,000	.0370″	3,092	291,000	?	?
" 2.............	170,000	.0430″	3,537	260,000	.0940″	15,097
" 3.............	130,000	.0278″	1,803	260,000	.1030″	16,867
" 4.	180,000	.0350″	3,495	280,000	.0990″	18,612
" 5.............	140,000	.0435″	2,617	250,000	.1130″	17,349
" 6.............	120,000	.0253″	1,580	251,000	.1090″	18,761
Average	151,670	.0353″	2,687	260,000	.1036″	17,337

NOTE.—The average resilience within the elastic limit of these piers was therefore about 15 per cent of their ultimate resilience.

The strength of brickwork varies considerably, according to the quality of brick and mortar used. Trautwine says that in some English experiments small cubical masses only 9 inches on each edge, laid in cement, crushed under from 27 to 40 tons per square foot. Some piers 9 inches square, 2′ 3″ high, set in cement and broken only two days after being built, required 44 to 62 tons per square foot to crush them. Another pier of pressed brick, in best Portland cement, was said to have withstood 202 tons per square foot, and with common lime mortar only one fourth as much.

In an article in *Engineering*, 1872, it is said that many hand-made, ill-burnt bricks will not stand more than a pressure of 14 tons per square foot, while an uncommonly strong machine-made brick by Clayton & Co. was found by Kirkaldy to sustain a pressure equal to 323 tons per square foot.

According to Robertson, piers 8½″ square, 2′ 6″ high, sustain 50 tons per square foot, when set in gray stone lime, and 200 tons per square foot, when set in Portland cement.

Clarke found that the resistance to crushing of rather soft brick set in cement averaged 34 tons; this seems to be considered by the writer of the article referred to to represent fairly the average resistance of ordinary stock bricks set in ordinary good mortar.

The Aide-Mémoire, Royal Engineers, gives also low figures for compressive strength of brickwork. For bricks set in mortar (meaning probably lime mortar), 20 tons per square foot is given; when set in cement, 30 tons.

In " Notes on Building Construction" we find for brick piers having a height of less than twelve times their least thickness :

Weight per square
foot at which
crushing commences.
Tons.

Bricks, hard stock, best quality, set in Portland cement and sand,
  I to I, 3 months old 40
Bricks, ordinary well-burnt, London stock, 3 months old........... 30
Bricks, hard stock, Roman cement and sand, 1 to 1, 3 months old.. 28
Bricks, hard stock, Lias lime and sand, 1 to 2, 6 months old........ 24
Bricks, hard stock, gray chalk lime and sand, 1 to 2, 6 months old.. 12

Some tests with piers of brickwork had been made at the Watertown Arsenal by direction of Colonel T. T. S. Laidley, Ordnance Department, United States Army, some time previous to those described in this report. The following table gives the results of those tests, from data obtained from the records at the arsenal. It is believed that these piers were about one year old when broken.

TABLE F₁.

COMPRESSIVE STRENGTH OF BRICK PIERS.

[From experiments made by direction of Col. T. T. S. Laidley, Ordnance Department, U.S.A.]

Nominal.	CROSS-SECTION. Actual.	CROSS-SECTION. Area. Square inches.	LENGTH. Inches.	LENGTH. Courses.	Weight, lbs.	Solid or Hollow	Kind of Brick.	MORTAR. Lime.	MORTAR. Cement.	MORTAR. Sand.	Strength of pier, pounds.	Strength per sq. ft., tons.
8″ sq.	7″.9 × 7″.9	62.4	80.05	34	386	Solid	Eastern	1	..	3	96,100	99.0
8″ "	....	57.8	....	7	74	"	Face—b	..	1	2	218,100	242.6
8″ "	7″.55×7″.55	57.0	16.125	7	73	"	?	1	..	3	143,600	162.0
8″ "	7″.8 × 7″.8	60.84	16.48	7	78.5	"	{ New Eastern }	1	..	3	148,400	156.8
12″ "	12″.1 × 12″.1	146.41	24.1	10	?	"	{ Old Bay State }	1	..	3	201,000	88.25
12″ "	{ 11″.5 × 11″.5 4″.25 × 4″.35 }	113.76	23.04	10	?	Hollow	Face—b	1	..	3	226,100	127.8
16″ "	15″.9 × 15″.9	252.8	....	13	?	Solid	{ New Eastern }	..	1	2	696,000	177.0

CHAPTER VIII.

SUMMARY.

In making the experiments which form the subject of these notes, it was not the intention to decide upon the relative merits, for building purposes, of the several kinds of material employed, but to obtain some further information (which could be secured only through the aid of the powerful testing-machine at the Watertown Arsenal) regarding the behavior under compressive stress of both natural and artificial stone in various gradations of size, from cubes of one or two inches on a side up to as large cubes as the machine was able to break. As stated in the opening remarks, the tests were practically a continuation of those made about twelve years ago, described in my report of August 10, 1875.

The results and conclusions may be summed up as follows:

1. As indicated by previous experiments, the interposition of wooden cushions in testing any material does not allow the full development of its compressive strength; the wood seems to induce or favor cleavage of the test-piece in a direction parallel to its fibres.

2. To secure uniformity of results, any material which cannot be brought to a satisfactorily smooth and plane surface on its bed-faces should receive a thin coating of some suitable substance: a film made with paste of plaster of Paris was found to answer very well.

3. The law of increase of compressive strength per square inch of bed-surface, with increasing size of cubes, which was based upon experiments made some ten years ago with various but limited sizes of Berea sandstone, was not confirmed when larger cubes of Haverstraw sandstone, cement, mortars, and concretes were tested. That some such law exists for cubes within certain limits cannot be doubted, not only in view of the Staten Island experiments, but of experiments made by

foreign investigators referred to in this report. The failure of
the law with larger cubes seems to be due to the lack of homogeneity throughout the mass of such pieces; this is indicated by
the strain-diagrams. It is only possible to discover defects in
a large piece by dividing it into smaller pieces; and when the
most perfect of these fragments are selected to prepare small
test-samples, approximately true units in regard to homogeneity of structure may be obtained. It is thought that
large cubes are not such units, or true monoliths; that they
represent a species of conglomerate of smaller irregular pieces,
bound together by a cementing substance of varying strength,
and perhaps partially separated by minute cracks and cavities.
With cements, mortars, and concretes, the relative amount of
work expended in consolidating the material in the moulds
cannot well be evenly distributed or proportioned for all sizes
of cubes; the amount of set developed in small and large
cubes of the same age is undoubtedly different. This is probably the reason why in all of the cements, mortars, and concretes the smallest sizes of each series of cubes proved the
strongest per square inch of surface pressed.

4. Since small cubes exhibited relatively the greatest compressive strength, while the material actually employed in
structures has much larger dimensions, the test-pieces should
preferably be made of larger-sized cubes in order to obtain
results of direct practical value.

5. That prisms of the same cross-section as cubes, but of
less height, are superior in strength to such cubes, has been
known before; the tests made at the Watertown Arsenal have
led to the construction of an empirical formula, expressing
the probable ratio of an increase of static strength as the
height of the prism is diminished.

6. The observations of compression, elasticity, and resilience are believed to form a contribution of some value toward
a better knowledge of the qualities and intrinsic merits of the
kinds of material tested. Little or nothing is found in print
on this subject. Information concerning the elasticity of
building material, especially of cement, and of concretes of
which such cement combined with sand forms the matrix,

cannot be otherwise than useful. Generally it is deemed sufficient to test the tensile strength of briquettes of cement, and when these can carry a certain load after a certain number of days, the cement is accepted. But there is not much known about its relative value when used in combination with sand, gravel, and broken stone. A large amount of scientific knowledge and skill has for many years past been applied to ascertain the properties of iron and steel, but very little attention has been paid to the subject of mortars and concretes. The importance of knowing whether such material possesses elasticity and resilience, and if so, to what extent, is very great, because structures are not merely subject to dead loads or statical strains; but also, in many cases, to live loads or dynamical strains. Masonry laid in cement or cement mortar, brickwork, and concrete, especially when used in foundations to support heavy moving machinery, are exposed to almost constant but ever-varying jar, vibration, and concussion.

In many instances such foundations have ultimately failed.

In an article in *The Engineer* of 1871 it was pointed out that the repeated failure of large engineering works, such as breakwaters, docks, walls, etc., is due, indirectly, to the want of elasticity of the cement used, and that for that reason it was necessary to know the extent to which cements, mortars, and concretes, possess the necessary quality of elasticity and resilience. This matter is of great importance in works of fortification where structures built of similar material, although covered with earth and sand, are exposed to violent concussion from the impact of heavy projectiles.

7. Further experiments in various directions seem to be desirable. Berea sandstone being, as far as tested upon a small scale, of exceptionally homogeneous structure, several sets of cubes might be procured, beginning with, say, 1-inch cubes, increasing very gradually in size to as large a cube as will call for the full strength of the most powerful available testing-machine.

Prisms of various material, both of less and greater height than corresponding cubes, and of various forms and sizes of

8

cross-sections, should be tested, singly as well as combined, both as dry-jointed and as cemented piers.

Experiments should be made to ascertain the ultimate compressive strength, elasticity, resilience, etc., of the best known and marketable cements, and of the mortars and concretes made with them. The same cements and mortars should simultaneously be tested as to their tensile strength.

Parallel tests should be carried on by repetition of loads below the crushing load in order to ascertain the existence of a law by which it may be possible to discover the maximum load which can alternately be put and taken off without injuring any given piece.

Finally, it would be well to try the effect of weights falling from certain heights upon material whose resistance, both under steady pressure carried to the crushing-point, and also under repeated loads, is known. In one series of tests the weight might be arranged to strike the entire surface of the bed, in another to strike a knife-edge blow, corresponding to the cutting edge of the face-hammer used in quarries.

APPENDIX.

GENERAL TABLE I.

COMPRESSIVE TESTS OF HAVERSTRAW FREESTONE (NEW YORK).

REMARKS.—*The Beds or Compressed Surfaces of the samples had been brought to a smooth surface by rubbing. Most of them were, moreover, thinly coated with plaster of Paris, which was allowed to harden before the test was made; but the samples marked * were not treated in this manner, but were rubbed once more to produce as smooth and even a surface as practicable.*

FORM AND NOMINAL SIZE.	Mark.	ACTUAL SIZE. Bed.	Of Sample.	Height. Including Plaster.	Weight of Sample.	CRUSHING STRENGTH IN POUNDS. Of Sample.	Per Square Inch.	1 er Cube Inch.	REMARKS.
1-inch Cube	a	1".00 × .98"	1".01	1".04	— lbs. 1.27 oz.	6,820	6,959	6,890	No preliminary signs of yielding.
1-inch Cube	b	1".00 × .98"	.98	1".02	— lbs. 1.23 oz.	6,960	7,102	7,395	No preliminary signs of yielding.
Average							7,030	7,142	
2-inch Cube	*a	2".04 × 2".04	2".01		— lbs. 10½ oz.	24,780	5,954	2,962	One pyramid well developed, its apex nearly reaching the opposite bed.
2-inch Cube	*b	2".03 × 2".04	2".04		— lbs. 11 oz.	23,800	5,747	2,871	One pyramid developed: its apex broken off.
Average							5,850	2,889	
2-inch Cube	c	2".04 × 2".05	2".04	2".06	— lbs. 10½ oz.	28,080	6,714	3,291	Two pyramids developed, appearing to slide obliquely past each other.
2-inch Cube	d	2".04 × 2".06	2".04	2".08	— lbs. 10½ oz.	23,480	5,587	2,739	Pyramids imperfect.
Average							6,150	3,015	

No preliminary signs of yielding.

GENERAL TABLE I.—(Continued.)

COMPRESSIVE TESTS OF HAVERSTRAW FREESTONE (NEW YORK).

Form and Nominal Size.	Mark.	Actual Size.			Weight of Sample.	Crushing Strength in Pounds.			Remarks.
		Bed.	Height Of Sample.	Including Plaster.		Of Sample.	Per Square Inch.	Per Cube Inch.	
3-inch Cube......	*a	2".97 × 2".99	2".94	......	2 lbs. ¾ oz.	54,900	6,182	2,103	Two rather pointed pyramids developed.
3-inch Cube......	*b	2".98 × 2".98	2".97	......	2 lbs. 1¼ oz.	51,300	5,799	1,953	One pyramid developed, the other rudimentary.
Average........		......	......	......	......	......	5,990	2,028	
3-inch Cube......	c	3".02 × 3".03	3".01	3".02	2 lbs. 1½ oz.	65,900	7,202	2,393	Two pyramids, appearing to pass each other obliquely.
3-inch Cube......	d	3".02 × 3".04	3".01	3".02	2 lbs. 1½ oz.	53,200	5,795	1,925	Good sample of two pyramids adhering to each other.
Average........		......	......	......	......	......	6,498	2,159	
4-inch Cube......	*a	3".99 × 4".00	3".99	......	5 lbs. 1 oz.	101,200	6,341	1,585	Bed slightly convex. } No preliminary signs of yielding.
4-inch Cube......	*b	3".98 × 3".99	3".98	......	5 lbs. ¾ oz.	84,800	5,340	1,341	Bed slightly convex.
Average........		......	......	......	......	......	5,840	1,463	
4-inch Cube......	c	4".05 × 4".05	4".01	4".05	5 lbs. 1¾ oz.	97,600	5,950	1,484	No preliminary signs of yielding.
4-inch Cube......	d	4".02 × 4".04	4".00	4".01	5 lbs. ½ oz.	100,900	6,213	1,553	No preliminary signs of yielding.
Average........		......	......	......	......	......	6,081	1,518	
5-inch Cube......	*a	4".96 × 5".00	4".95	......	9 lbs. 11 oz.	141,100	5,658	1,148	Cracked at 131,000 lbs. One well-developed pyramid.
5-inch Cube......	*b	4".94 × 4".96	5".00	......	9 lbs. 11 oz.	129,900	5,301	1,060	No preliminary signs of yielding. Two irregular pyramids.
Average........		......	......	......	......	......	5,479	1,104	

									Remarks
5-inch Cube......	c	4".98 × 5".05	4".98	5".01	9 lbs. 9¾ oz.	202,500	8,052	1,617	First crack at 200,100 lbs. Two pyramids of about equal size. Two sides of the cube came off almost entire.
5-inch Cube......	d	4".94 × 5".04	4".97	5".00	9 lbs. 10½ oz.	170,000	6,828	1,374	No preliminary signs of yielding. Two well-developed pyramids.
Average......							7,440	1,495	
6-inch Cube......	†a	5".90 × 5".90	5".88	5".99	16 lbs. 8½ oz.	249,900	7,179	1,221	No preliminary cracking. Two pyramids, finely developed. One side of the cube fitted into the angular recess formed by the pyramids, with scarcely any powdered material interposed.
6-inch Cube......	b	5".94 × 5".92	5".94	6".02	16 lbs. 4½ oz.	246,600	7,048	1,287	No preliminary cracking. The apex of one pyramid was found buried in the crater-like, mutilated apex of the other pyramid.
6-inch Cube......	c	5".94 × 5".94	5".95	6".05	16 lbs. 6½ oz.	262,300	7,147?	1,256	First crack at 259,000 lbs. One large pyramid well developed, the other small and imperfect.
6-inch Cube... ..	d	5".94 × 5".95	5".95	6".04	16 lbs. 9½ oz.	272,800	7,219	1,279	No preliminary cracking. Two pyramids of about equal size, not well developed.
Average......						......	7,334	1,236	NOTE.—In these four 6-inch cubes the sides separated finely from the pyramids.

† One bed of this cube was experimentally faced by means of some patent mechanical contrivance, which did not prove sufficiently successful, and the bed was subsequently replastered.

GENERAL TABLE I.—(*Continued.*)
COMPRESSIVE TESTS OF HAVERSTRAW FREESTONE (NEW YORK)

FORM AND NOMINAL SIZE.	Mark.	ACTUAL SIZE. Bed.	Height. Of Sample.	Height. Including Plaster.	Weight of Sample.	CRUSHING STRENGTH IN POUNDS. Of Sample.	Per Square Inch.	Per Cube Inch.	REMARKS.
7-Inch Cube......	a	7″.04 × 7″.08	7″.00	7″.07	27 lbs. — oz.	304,800	6,115	874	One pyramid well developed; its apex fitted into a corresponding cavity of the opposite pyramid.
7-inch Cube......	b	7″.03 × 7″.04	7″.02	7″.10	27 lbs. — oz.	283,500	5,728	815	Pyramids same as above. Of the four sides of the cube, one came off entire, another nearly so, and the rest in large fragments.
7-inch Cube......	c	7″.03 × 7″.05	6″.96	7″.02	27 lbs. 8 oz.	326,600	6,590	947	Two pyramids; the whole cube shattered.
7-inch Cube......	d	6″.97 × 7″.00	7″.00	7″.10	27 lbs. — oz.	302,000	6,190	884	Two pyramids; the whole cube shattered.
Average.........							6,156	880	
8-inch Cube......	a	7″.99 × 7″.99	7″.99	8″.15	39 lbs. — oz.	397,000	6,219	778	Two irregular pyramids. One of the cube broke off of the lateral faces entire; the others in fragments.
8-inch Cube......	b	8″.05 × 8″.16	8″.00	8″.14	41 lbs. 4 oz.	438,400	6,674	834	Two pyramids, one side steep (about 80°), the other about 45°.
8-inch Cube......	c	8″.00 × 8″.03	8″.00	8″.07	39 lbs. 12 oz.	388,000	6,040	755	Same as 8-inch cube b.
8-inch Cube......	d	8″.02 × 8″.02	7″.96	8″.04	39 lbs. 8 oz.	395,700	6,152	773	
Average.........							6,271	785	

No preliminary signs of yielding.

Specimen					Weight	Load			Remarks
9-inch Cube	a	8".99 × 9".07	8".96	9".05	56 lbs. — oz.	470,400	5,769	644	No preliminary cracking. One large pyramid well developed, and one small pyramid irregular
9-inch Cube	b	9".00 × 9".03	8".97	9".02	57 lbs. 12 oz.	558,000	6,989	779	First crack at 536,000 lbs. Two pyramids, rather well developed. Lateral faces shattered.
9-inch Cube	c	9".02 × 9".04	9".01	9".05	57 lbs. 12 oz.	643,000	7,886	875	No preliminary cracking. Results of fracture, same as in preceding cube.
9-inch Cube	d	8".99 × 9".01	8".92	8".99	56 lbs. 8 oz.	415,000	5,494	616	No preliminary cracking. Results of fracture, same as in preceding cube.
Average							6,534	728	
10-inch Cube	a	9".96 × 10".02	10".01	10".07	79 lbs. 12 oz.	520,000	5,210	520	First crack at 500,000 lbs.
10-inch Cube	b	9".80 × 10".00	10".01	10".12	77 lbs. 8 oz.	650,500	6,638	663	No preliminary cracking. After fracture a layer of comparatively coarse material was found in the body of the cube, running across it, about parallel to the bed-faces.
10-inch Cube	c	9".96 × 10".00	10".01	?	78 lbs. 4 oz.	840,000?	8,434?	842?	Cracked at the maximum load of 800,000 lbs. The pressure was then reduced to 5000 lbs., and gradually raised again to the maximum load, which failed to crush the cube. In the table it is arbitrarily assumed that a load of 840,000 lbs. would have broken it.
10-inch Cube	d	9".98 × 10".00	10".00	10".09	78 lbs. 4 oz.	644,000	6,446	645	No preliminary signs of yielding.
Average							6,682?	668?	
11-inch Cube	a	11".00 × 11".05	10".92	11".09	105 lbs. — oz.	792,000	6,508	596	First crack at 770,000 lbs. One pyramid well formed, the other shattered.
11-inch Cube	b	10".96 × 11".10	11".01	11".08	106 lbs. 8 oz.	785,000	6,453	586	First crack at 770,000 lbs. One pyramid well formed, the other shattered.
11-inch Cube	c	11".00 × 11".00	10".97	11".01	104 lbs. 8 oz.	779,200	6,440	587	Cracked at 778,000 lbs. Two rather well-formed pyramids; balance of stone shivered to pieces.
11-inch Cube	d	11".05 × 11".10	11".02	11".16	106 lbs. 8 oz.	769,000	6,270	569	No preliminary signs of yielding.
Average							6,418	584	

GENERAL TABLE I.—(Continued.)

COMPRESSIVE TESTS OF HAVERSTRAW FREESTONE (NEW YORK).

FORM AND NOMINAL SIZE.	Mark.	ACTUAL SIZE.			Weight of Samples.	CRUSHING STRENGTH IN POUNDS.			REMARKS.
		Bed.	Of Sample.	Includ- ing Plaster.		Of Sample.	Per Square Inch.	Per Cube Inch.	
			Height.						
12-inch Cube.....	a	11".95 × 12".00	12".01	12".05	139 lbs. 8 oz.	?	?	?	Not broken under maximum load of 800,000 lbs.
12-inch Cube.....	b	12".00 × 12".00	12".04	12".23	138 lbs. — oz.	?	?	?	Not broken under maximum load of 800,000 lbs.
12-inch Cube.....	c	11".96 × 12".00	12".00	12".20	135 lbs 8 oz.	764,000	5,323	444	First crack at 700,000 lbs.
12-inch Cube.....	d	11".90 × 11".96	12".01	12".14	135 lbs. 12 oz.	?	?	?	Not broken under maximum load of 800,000 lbs
Average..									

NOTE.—All the freestone cubes enumerated in Table I. that were actually broken burst with a dull explosive sound. Of those samples that were not crushed under the maximum load of 800,000 pounds, 10-inch cube c was subsequently tested in conjunction with three prismatic slabs of neat Dyckerhoff Portland cement, each measuring 11" × 12" × 2". See report. The three refractory 12-inch freestone cubes, a, b, and d, were subsequently combined as a dry-jointed pier, and broken in that form. For the results of this test see end part of Special Table I. and report.

PRISMS.									
4" × 4" × 1".....	a	3".99 × 4".00	1".01	1".09	1 lb. 4½ oz.	299,800	18,784	18,598	Crackling noise heard some time before fracture occurred. Sides well shattered and partly ground to powder; remainder of little cohesive strength.
4" × 4" × 1".....	b	3".98 × 4".00	1".03	1".11	1 lb. 5 oz.	224,000	14,071	13,660	First crack at 142,000 lbs. The prism was entirely disinte- grated.
Average......						16,427	16,129		

		Dimensions			Weight				Remarks
4″ × 4″ × 2″	a	3″.97 × 3″.98	1″.98	2″.07	2 lbs. 7½ oz.	129,000	8,221	4,152	First crack at 122,000 lbs. Remaining core in the form of two combined frustated pyramids.
4″ × 4″ × 2″	b	4″.00 × 4″.03	2″.02	2″.11	2 lbs. 8½ oz.	126,300	7,835	3,879	First crack at 116,000 lbs. The lateral faces separated in the form of slabs, varying in size from 2″× 1″.5 to 2″×3″.5.
Average						....	8,028	4,015	
4″ × 4″ × 3″	a	4″.00 × 4″.03	3″.04	3″.11	3 lbs. 14 oz.	99,000	6,141	2,020	First crack at 83,200 lbs. Two rather well-formed pyramids which separated when the lateral fragments were removed. Apex of one pyramid was pointed; that of the other had the form of a sharp ridge.
4″ × 4″ × 3″	b	4″.00 × 4″.05	3″.02	3″.14	3 lbs. 13½ oz.	116,900	7,216	2,389	First crack at 113,000 lbs. Two connected frustated pyramids. Sides of prism came off in the shape of good-sized slabs.
Average						....	6,698	2,204	
8″ × 8″ × 2″	a	8″.05 × 8″.08	2″.07	2″.20	10 lbs. 9½ oz.	800,000 +	12,300 +	5,942 +	First crack at 573,000 lbs. When the maximum load of 800,000 lbs. had been reached, some lateral spawls only had broken off, leaving a solid prism about 7¾ inches square. It is estimated that with an additional load of 30,000 lbs. the piece might have been broken.
8″ × 8″ × 2″	b	7″.95 × 7″.98	2″.02	2″.14	10 lbs. 4 oz.	800,000 +	12,610 +	6,243 +	First crack at 640,000 lbs. When the maximum load of 800,000 lbs. had been removed the prism was found to be but little injured.
Average						....	12,455 +	6,092 +	

GENERAL TABLE I.—(*Continued.*):
COMPRESSIVE TESTS OF HAVERSTRAW FREESTONE (NEW YORK).

Form and Nominal Size.	Mark.	Actual Size.			Weight of Sample.	Crushing Strength in Pounds.			Remarks.
		Bed.	Of Sample.	Height. Including Plaster.		Of Sample.	Per Square Inch.	Per Cube Inch.	
8″ × 8″ × 3″	a	7″.95 × 8″.03	3″.04	3″.19	15 lbs. 8½ oz.	658,500	10,251	3,372	First crack at 512,000 lbs. The prism was well disintegrated. When the lateral fragments and stone rubbish had been removed, two small frustated, irregular pyramids remained.
8″ × 8″ × 3″	b	8″.00 × 8″.09	3″.08	3″.27	15 lbs. 11 oz.	571,000	8,823	2,864	First crack at 512,000 lbs. When the outer lateral fragments had been removed, two irregular frustated pyramids appeared. On continuing removing fragments from these, other pyramids of smaller size were developed, until two rather solid pyramids remained, with bases of about 4″.5 × 5″, and 5″ × 5″, respectively.
Average.........							9,537	3,118	
8″ × 8″ × 4″.........	a	8″.05 × 8″.05	4″.03	4″.18	20 lbs. 6 oz.	695,000	9,336	2,317	First crack at 579,000 lbs. First crack at 454,000 lbs. { In each case the prism was well broken up, two small, badly shaped pyramids remaining, separated from each other.
8″ × 8″ × 4″	b	8″.08 × 8″.08	4″.07	4″.18	21 lbs. 4 oz.	907,000	7,766	1,908	
Average.........							8,551	2,112	

									Remarks
8" × 8" × 5"........	a	7".97 × 8".04	5".00	5".16	24 lbs. 7¼ oz.	444,500	6,937	1,387	First crack at 370,000 lbs. Well broken up; two small pyramids, incompletely developed.
8" × 8" × 5"........	b	8".10 × 8".15	5".03	5".15	26 lbs. 8 oz.	567,000	8,591	1,708	No preliminary cracking. Well broken up; two small pyramids, incompletely developed.
Average........		....	....	....	....	....	7,764	1,547	
8" × 8" × 6"........	a	8".04 × 8".08	5".92	6".03	29 lbs. 11 oz.	408,000	6,281	1,061	{In each case, two rather regular pyramids remained, otherwise the sample was well broken up.} No preliminary cracking. First crack at 409,000 lbs.
8" × 8" × 6"........	b	7".90 × 8".00	5".98	6".09	29 lbs. 7 oz.	429,000	6,383	1,135	
Average........		....	....	....	....	....	6,534	1,098	
8" × 8" × 7"........	a	7".95 × 7".97	7".00	7".09	35 lbs. — oz.	423,800	6,689	936	First crack at 403,000 lbs. No preliminary signs of yielding.
8" × 8" × 7"........	b	8".00 × 8".05	6".90	7".00	34 lbs. 8 oz.	421,000	6,537	947	
Average........		....	....	....	....	....	6,613	951	

GENERAL TABLE II.

NEAT CEMENT.

COMPRESSIVE TESTS OF DYCKERHOFF'S PORTLAND CEMENT (GERMANY).

NOMINAL SIZE.	Mark.	ACTUAL SIZE. Bed.	Height.	How Broken.	Age when Crushed. (y. m. d.)	Weight of Sample. (lbs. oz.)	CRUSHING STRENGTH— Of Sample. (lbs.)	Per Square Inch. (lbs.)	Per Cubic Inch. (lbs.)	REMARKS.
1-inch Cube.	a	1".03 x 1".98	1".01	Beds plast'r'd	1 10 28	1.20	5,710	5,657	5,601	The aggregate thickness of plaster on the two end-faces varied from .005 inch to .05 inch. The pyramidal formation, after fracture, was incompletely developed, but manifest. The sides of the cubes generally separated well.
" "	b	1".00 x 1".02	1".01	"	1 10 28	1.21	6,050	5,931	5,873	
" "	c	1".02 x 0".98	1".00	"	1 10 28	1.18	5,000	5,902	5,902	
" "	d	1".02 x 1".02	0".98	"	1 10 28	1.20	5,880	5,652	5,767	
" "	e	1".01 x 1".00	1".02	"	1 10 28	1.21	6,120	6,059	5,904	
" "	f	1".00 x 1".02	1".02	"	1 10 28	1.23	6,300	6,176	6,055	
Average...								5,896	5,850	
2-inch Cube.	a	2".02 x 2".03	2".03	Directly.	1 10 3	10.00	33,300	8,121	4,000	Cracked at 17,900 lbs.
" "	b	2".00 x 1".95	2".00	"	1 10 3	9.4	25,450	6,525	3,263	
" "	c	2".00 x 1".99	2".00	"	1 10 3	9.5	24,400	6,130	3,065	Corner cracked off.
" "	d	2".03 x 2".00	2".01	"	1 10 3	9.75	29,480	7,261	3,612	Snapping sound at 22,100 lbs.
" "	e	2".01 x 1".98	2".02	"	1 10 3	9.62	25,100	6,307	3,122	First crack at 28,100 lbs.
" "	f	2".03 x 1".99	2".04	Beds plast'r'd	1 10 28	9.75	33,000	8,218	4,099	
Average...								7,094	3,848	

Mark	Ref.	Dimensions of Beds	Height	Bearing					Breaking Weight, lbs.	lbs. per sq. in.		Remarks
3-inch Cube	a	2".98 × 2".96	2".96	Directly	1	10	3	15¾	52,900	5,997	2,026	First cracking sound at 36,800 lbs.
3 "	b	2".98 × 2".95	2".97	"	1	10	3	15½	47,920	5,578	1,886	Cracked at 19,000 lbs., one side split off.
3 "	c	2".97 × 2".94	2".95	"	1	10	3	15	50,400	5,772	1,948	Cracked at 19,200 lbs.
3 "	d	2".98 × 2".93	2".96	"	1	10	3	15½	48,400	5,634	1,993	No cracks observed previous to fracture.
3 "	e	2".98 × 2".98	2".95	Beds plast'r'd	1	10	3	15½	51,900	5,840	1,981	
3 "	f	2".98 × 2".97	2".97	"	1	10	27	15¼	60,000	6,795	2,288	
Average										5,937	2,005	
4-inch Cube	a	4".00 × 4".00	3".98	Directly	1	10	3	11	82,200	5,138	1,292	Snapping sounds at 72,000 lbs.
4 "	b	4".05 × 4".05	4".02	"	1	10	4	15	88,500	5,395	1,342	First crack at 54,800 lbs.
4 "	c	3".97 × 3".98	3".97	"	1	10	4	15	66,500	4,333	1,092	Snapping at 49,000 lbs. Uneven bearing of bed-faces.
4 "	d	4".07 × 4".05	4".02	"	1	10	5	—	99,350	5,481	1,364	Two well-developed pyramids, bases nearly sizes of beds.
4 "	e	4".02 × 3".99	4".02	"	1	10	4	13½	74,000	4,642	1,148	Pressed surfaces slightly round.
4 "	f	4".02 × 4".00	4".02	"	1	10	4	13¼	66,300	4,173	1,026	
Average										4,847	1,211	
5-inch Cube	a	5".02 × 5".02	5".01	Directly	1	10	9	7	104,500	4,145	828	At 88,700 lbs. corner flaked off.
5 "	b	5".01 × 5".07	5".00	"	1	10	9	8	116,700	4,594	919	At 110,700 lbs., snapping sounds.
5 "	c	5".08 × 5".03	5".02	"	1	10	9	10	125,900	4,927	981	At 100,800 lbs., cracked along one corner.
5 "	d	5".01 × 5".00	5".01	"	1	10	9	6¼	119,900	4,786	955	At 101,000 lbs., cracked along one corner.
5 "	e	4".92 × 4".90	5".00	"	1	10	9	4	121,500	5,040	1,003	At 96,500 lbs. cracked at corner.
5 "	f	4".96 × 4".97	5".01	"	1	10	9	4	102,800	4,170	832	Corner not taking bearing.
Average										4,610	920	
6-inch Cube	a	6".00 × 6".00	5".98	Directly	1	10	16	2	143,000	3,972	664	At 90,000 lbs., upper side cracked.
6 "	b	6".00 × 6".00	5".92	"	1	10	16	—	129,000	3,582	605	Cracked at 111,000 lbs.
6 "	c	6".07 × 6".00	5".97	"	1	10	16	10	160,300	4,401	737	Cracked at 137,000 lbs.
6 "	d	6".02 × 6".00	5".97	"	1	10	16	3	179,700	4,975	833	
6 "	e	5".98 × 6".00	5".97	"	1	10	16	—	133,000	3,762	630	Cracked at 130,000 lbs.
6 "	f	5".98 × 5".99	5".98	"	1	10	16	1	176,500	5,093	836	Cracked at 174,000 lbs.
Average										4,283	717	

GENERAL TABLE II.—(*Continued.*)

NEAT CEMENT.

COMPRESSIVE TESTS OF DYCKERHOFF'S PORTLAND CEMENT (GERMANY).

NOMINAL SIZE.	Mark.	ACTUAL SIZE. Bed.	Height.	How Broken.	Age when Crushed. y. m. d.	Weight of Sample. lbs. oz.	CRUSHING STRENGTH— Of Sample. lbs.	Per Square Inch. lbs.	Per Cubic Inch. lbs.	REMARKS.
7-inch Cube.	a	7″.07 × 7″.11	6″.99	Directly.	1 10 4	26 6	228,900	4,554	651	Cracked at 64,000 lbs. Two corners not taking full bearing.
7 " "	b	7″.07 × 7″.08	7″.00	"	1 10 4	26 4	192,700	3,849	550	Cracked at 158,000 lbs.
7 " "	c	7″.09 × 7″.14	7″.05	"	1 10 4	26 12	259,000	5,134	728	At 188,000 lbs, cracked along bottom side, central part.
7 " "	d	7″.11 × 7″.02	7″.04	"	1 10 5	26 6	288,000	5,774	820	Cracked at 232,000 lbs. Two pyramids appearing to slide one upon the other.
7 " "	e	7″.11 × 7″.20	7″.02	"	1 10 5	26 15	265,200	5,180	733	
7 " "	f	7″.02 × 7″.11	7″.01	"	1 10 5	26 4	271,000	5,429	774	
Average.								4,987	710	
8-inch Cube.	a	8″.00 × 8″.05	8″.00	Directly.	1 10 5	38 10	289,000	4,488	361	Cracked at 266,000 lbs.
8 " "	b	8″.03 × 8″.10	7″.99	"	1 10 5	37 8	301,100	4,609	579	Cracked at 238,000 lbs.
8 " "	c	8″.03 × 8″.07	8″.00	"	1 10 5	37 8	294,100	4,540	567	Snapping sound at 180,000 lbs.
8 " "	d	8″.04 × 8″.00	8″.04	"	1 10 5	39 —	360,000	5,597	696	At 250,000 lbs, begins to scale off.
8 " "	e	7″.98 × 8″.03	8″.02	"	1 10 5	38 —	299,200	5,533	690	At 296,000 lbs, broke at corner.
8 " "	f	8″.00 × 8″.04	8″.00	"	1 10 5	39 —	338,000	5,255	657	Cracked at 304,000 lbs.
Average.								5,007	625	

		Dimensions	Height	Bedding					Load, lbs.	Per sq. in.	Ratio	Remarks
9-inch Cube	a	9″.05 × 9″.01	9″.04	Directly	10	4	56	—	373,000	4,574	506	Begins to crack at 345,000 lbs.
9 ″	b	9″.02 × 9″.12	9″.05	″	11	4	56	—	373,000	4,594	508	Corner off at 327,000 lbs.
9 ″	c	9″.00 × 9″.00	8″.99	″	11	4	55	—	396,000	4,689	544	At 330,000 lbs. cracked at corner. Yielded suddenly at 395,000 lbs.
9 ″	d	9″.02 × 9″.04	9″.05	″	10	4	56	8	390,000	4,783	528	Burst suddenly at 390,000 lbs. Slight crack at 350,000 lbs.
9 ″	e	9″.07 × 9″.00	9″.03	″	11	5	56	—	468,200	5,736	635	Begins to crack at 458,000 lbs.
9 ″	f	9″.05 × 9″.10	8″.98	″	11	5	55	8	375,000	3,946	439	Cracked at 130,000 lbs.
Average										4,754	527	
10-inch Cube	a	10″.08 × 10″.05	9″.97	Directly	10	5	75	8	395,300	3,902	391	At 318,000 lbs. cracked at bottom. More cracking at 340,000 lbs.
10 ″	b	10″.10 × 10″.00	10″.00	″	10	5	76	8	587,100	5,859	586	Cracked at 540,000 lbs.
10 ″	c	10″.00 × 10″.04	10″.00	″	10	5	76	8	519,000	5,123	512	At about 330,000 lbs. cracked along one edge.
10 ″	d	10″.08 × 10″.10	10″.00	″	10	5	77	—	429,100	4,225	422	
10 ″	e	10″.00 × 10″.05	10″.08	″	10	5	76	—	473,400	4,710	467	At 242,000 lbs. cracked on lower side.
10 ″	f	10″.01 × 10″.05	9″.99	″	10	5	76	8	477,600	4,747	475	
Average										4,761	475	
11-inch Cube	a	11″.00 × 11″.15	11″.00	Directly	10	5	101	—	591,200	4,820	435	Cracked at 530,000 lbs.
11 ″	b	11″.05 × 11″.00	11″.00	Beds plast'r'd	10	24	100	—	633,000	5,208	473	At 570,000 lbs. corner cracked.
11 ″	c	11″.00 × 11″.18	11″.00	″	10	24	101	8	725,000	5,835	526	No premonitory cracks.
11 ″	d	11″.01 × 11″.21	11″.00	″	10	24	101	8	574,000	5,451	466	Cracked at 528,000 lbs.
11 ″	e	11″.02 × 11″.21	10″.99	″	10	24	101	—	690,000	5,585	508	Cracks in sight at 680,000 lbs.
11 ″	f	11″.05 × 11″.05	11″.02	″	10	24	100	—	645,600	5,287	480	Cracks in sight at 635,000 lbs.
Average										5,374	488	
12-inch Cube	a	12″.05 × 12″.00	12″.00	Beds plast'r'd	10	24	129	—	710,000	4,910	409	Cracks in sight at 660,000 lbs.
12 ″	b	12″.08 × 12″.05	11″.97	″	10	24	129	—	783,000	5,379	449	Cracks in sight at 673,000 lbs.
12 ″	c	12″.00 × 12″.03	12″.03	″	10	24	130	8	800,000	?	?	Load of 800,000 lbs. repeatedly applied. See Special Table II.
12 ″	d	12″.00 × 11″.30	12″.00	″	10	26	123	—	800,000	?	?	Small pieces flew off at 798,000 lbs. Load of 800,000 lbs. several times applied. See Special Table II.
12 ″	e	12″.05 × 12″.00	12″.00	″	10	26	131	—	800,000	5,532	461	Small pieces flew off at 770,000 lbs. The sample failed rapidly when the load of 800,000 lbs. had been sustained ¼ minute.
12 ″	f	12″.00 × 12″.06	12″.00	″	10	26	130	—	773,200	5,343	445	
Average												

GENERAL TABLE II.—(Continued.)

NEAT CEMENT.

COMPRESSIVE TESTS OF DYCKERHOFF'S PORTLAND CEMENT (GERMANY).

Nominal Size.	Mark.	Actual Size.		How Broken.	Age when Crushed.	Weight of Sample.	Crushing Strength—			Remarks.
		Bed.	Height.		y. m. d.	lbs. oz.	Of Sample.	Per Square Inch.	Per Cubic Inch.	
PRISMS. 4"×4"×1"	a	4".03×3".98	1".03	Beds plast'r'd	1 10 26½	1 3½	lbs. 250,000	lbs. 15,587	lbs. 15,132	Numerous light crackling sounds, beginning with a load of 140,000 lbs., were heard during the process. At 250,000 lbs. pressure the piece was removed; the four sides of the prisms could easily be removed, but the remaining mass appeared sound. An initial crack was seen on one of the compressed surfaces; a slight blow with a hammer separated the prism in two pieces.
4"×4"×1"	b	4".05×4".04	1".02	"	1 10 26½	1 3½	275,000	16,807	16,641	The sides of the prism began to crack from 40,000 lbs. upwards; crackling sounds were heard from time to time. At 275,000 lbs. the piece was taken from the press; the sides and ground fragments being removed, about one half of the mass remained as a core, the substance of which appeared well disintegrated.
4"×4"×1"	c	3".98×4".06	1".02	"	1 10 26½	1 3½	275,000	17,018	16,685	Crackling sounds heard at loads of 100,000 lbs. and 125,000 lbs. respectively. When removed the prism was found to be well disintegrated, leaving only a core about 2¾"×2½" in cross-section, as a whole, cracked at several places.
Average								16,471	16,153	

4"×4"×2"	a	4".01 × 4".02	2".00	Beds plast'r'd	1	10	26	2	5½	102,100	6,334	3,168	Distinct crackling sounds from 40,000 lbs. upwards. The prism was well disintegrated; the remaining core was easily broken up by hand.
4"×4"×2"	b	4".05 × 4".01	2".03	"	1	10	26	2	7	104,200	6,416	3,151	First cracks at 78,000 lbs. Well disintegrated. Traces of pyramidical formation of central core, pressed faces as bases.
4"×4"×2"	c	4".06 × 4".02	2".02	"	1	10	26	2	7	103,800	6,360	3,148	First crack at 89,000 lbs. Same phenomena as before.
Average											6,370	3,159	
4"×4"×3"	a	4".04 × 4".02	3".00	Beds plast'r'd	1	10	26	3	5¾	87,000	5,357	1,786	Two frustated pyramids, well jointed. First crack at 68,000 lbs.
4"×4"×3"	b	3".99 × 3".99	3".02	"	1	10	29	3	9¼	95,600	6,005	1,988	Two frustated pyramids, well jointed. No preliminary cracking.
4"×4"×3"	c	4".02 × 4".02	3".02	"	1	10	26	3	10½	107,400	6,646	2,201	First crack at 97,000 lbs. Two frustated pyramids remained firmly connected, one much smaller than the other, its base on the corresponding surface being only about 2½ inches square, while the base of the larger pyramid was nearly of the size of the original bed-face of the prism.
Average											6,003	1,992	

GENERAL TABLE II.—(Continued.)

NEAT CEMENT.

COMPRESSIVE TESTS OF DYCKERHOFF'S PORTLAND CEMENT (GERMANY).

NOMINAL SIZE.	Mark	ACTUAL SIZE.		How Broken.	Age when Crushed.	Weight of Sample.	CRUSHING STRENGTH—			REMARKS.
		Bed.	Height.		y. m. d.	lbs. oz.	Of Sample. lbs.	Per Square Inch. lbs.	Per Cubic Inch. lbs.	
PRISMS. 8" x 8" x 2"	a	8".07 x 8".07	2".02	Beds plast'r'd	1 10 26	9 13	654,000	10,042	4,971	First crack at 283,000 lbs. When the sample was removed from the press, the outer portion was found thoroughly broken up; the core was apparently solid, but could easily be broken up in small pieces with light blows of a hammer.
8" x 8" x 2"	b	7".67 x 7".67	2".03	"	1 10 26	9 6	725,000	11,286	5,560	First cracking sound at 130,000 lbs. Cracked at 380,000 lbs. A double frustated pyramid remained, base about 5½" x 6½"; the lateral material shattered to pieces. The pyramidal core seemed to be solid, but on examination was found to have numerous cracks.
Average								10,664	5,265	
	c	8".01 x 4".76	2".01	Beds plast'r'd	1 11 00	5 11¼	317,500	8,327	4,143	First crack at 70,000 lbs. This sample was originally of the same size as a or b, but was accidentally dropped when being put in the machine, breaking in three pieces. The double pyramidal formation was well developed.

											Remarks
8"×8"×3" a	8".15×8".04	3".00	Beds plast'r'd	1	10	26	14	6	579,300	8,841	2,947	Commenced cracking at 313,-000 lbs. Piece well demolished; sides off; central portions slightly adhering together, but easily broken up by hand.
8"×8"×3" b	8".06×8".00	3".01	"	1	10	26	14	5	417,000	6,467	2,148	Cracked with 298,000 lbs. After removing the shattered fragments forming the sides a double frustated pyramid remained, bases about 5 inches square.
8"×8"×3" c	8".09×3".05	3".01	"	1	10	26	14	10½	407,000	6,230	2,076	First crack at 282,000 lbs. Result as with b, except that the bases of the pyramids were somewhat smaller.
Average										7,186	2,399	
8"×8"×4" a	8".06×8".04	4".01	Beds plast'r'd	1	10	28	19	3	394,800	6,092	1,519	Cracking at 242,000 lbs. Distinct formation of two pyramids with irregular bases, measuring about 7¾"×6" and 7"×6" respectively: oblique axis.
8"×8"×4" b	8".02×8".05	4".00	"	1	10	28	19	4	390,600	6,050	1,512	Cracking at 313,000 lbs. Two oblique frustated pyramids, bases of each about 6"×6½".
8"×8"×4" c	8".10×8".10	4".03	"	1	10	28	19	8	374,000	5,714	1,417	Cracking at 305,000 lbs. Two pyramids, axis nearly vertical.
Average										5,952	1,483	
8"×8"×5" a	8".00×8".02	5".09	Beds plast'r'd	1	10	28	24	7½	421,000	6,562	1,299	No preliminary cracking. Fair example of two frustated pyramids, axis nearly vertical.
8"×8"×5" b	8".05×8".12	5".04	"	1	10	28	24		388,000	5,844	1,160	Cracking at 310,000 lbs. Two pyramids developed, axis somewhat oblique.
8"×8"×5" c	8".05×8".00	5".06	"	1	10	28	24	4	364,000	5,652	1,117	Cracked at 296,000 lbs. Sample well broken up, pyramids but poorly developed.
Average										6,019	1,169	

GENERAL TABLE II.—(Continued.)

NEAT CEMENT.

COMPRESSIVE TESTS OF DYCKERHOFF'S PORTLAND CEMENT (GERMANY).

NOMINAL SIZE.	Mark.	ACTUAL SIZE.		How Broken.	Age when Crushed.	Weight of Sample.	CRUSHING STRENGTH—			REMARKS.
		Bed.	Height.		y. m. d.	lbs. oz.	Of Sample.	Per Square Inch.	Per Cubic Inch.	
							lbs.	lbs.	lbs.	
PRISMS.										
8″×8″×6″	a	8″.19×8″.14	6″.01	Beds plast'r'd	1 10 28	29 4	355,300	5,329	887	Cracking at 326,000 lbs. Formation of two pyramids, resembling in development those of a full cube.
8″×8″×6″	b	8″.02×8″.06	6″.04	"	1 10 28	28 15	396,100	6,128	1,015	Cracking at 365,800 lbs. A comparatively tough sample. The angular mass adhering to the slanting sides of the pyramids, separated but incompletely from them under repeated blows of a hammer. The pyramids were not well developed. It seemed as if a greater pressure should have been applied to disintegrate the piece to such a degree as to produce the usual phenomena.
8″×8″×6″	c	8″.12×7″.97	5″.96	"	1 10 28	26 9	379,000	5,856	983	Cracked at 373,000 lbs.
Average......								5,771	962	

Size		Aggregate										Remarks
13" x 12" x 2"	a	11".95 x 12".00	2".05	Beds plast'r'd	10	28	22	5	?	?	?	Cracking sound at 650,000 lbs. Not yielding under the maximum load of 800,000 lbs. applied for some time. b and c were not tested singly, as it was not expected that they could be broken. The prisms a, b, and c were then placed together so, as to form a dry-jointed pile, as follows: Not broken under the maximum load of 800,000 lbs.
12" x 12" x 2"	b	11".95 x 12".03	2".05	"	10	28	22	10	?	?	?	
12" x 12" x 2"	c	12".00 x 11".96	2".02	"	10	28	22	2	?	?	?	
12" x 12" x 2"	a, b, c	Aggregate >	6'.20	"	1	1	1	67	1	?	?	
12" x 12" x 4"	a	11".99 x 12".12	4".01	Beds plast'r'd	1	11	1	128	662,000	?	?	Prism a was first tried singly. A snapping sound was heard at 754,000 lbs. The piece resisted the ultimate load of 800,000 lbs. Prisms a, b, and c were then combined in a pile and so tested. The first crack was heard at 590,000 lbs.
12" x 12" x 4"	b	11".98 x 11".98	4".04	"								
12" x 12" x 4"	c	11".95 x 12".05	4".07	"								
12" x 12" x 6"	a	12".01 x 12".04	5".98	Beds plast'r'd					700,000			These three prisms were at once tested in combination as a pile or pier; a was next to the driving head, and c next to the fixed head. Snapping sounds were heard at 660,000 lbs. At 700,000 lbs. crack opened at a and b. Relaxing the load down to 5000 lbs. and gradually increasing it again, the piece rapidly failed at 690,000 lbs., and was crushed. General shape of fragments pyramidal, with steep sides at the outer prisms, the line of fracture extending into the middle prism; the latter was the one most seriously shattered.
13" x 12" x 6"	b	12".05 x 11".99	5".94	"	1	10	28	194	2	?		
12" x 12" x 6"	c	12".13 x 12".68	5".95	"								

GENERAL TABLE II.—(Continued.)

NEAT CEMENT.

COMPRESSIVE TESTS OF DYCKERHOFF'S PORTLAND CEMENT (GERMANY).

NOMINAL SIZE.	Mark.	ACTUAL SIZE.		How Broken.	Age when Crushed.	Weight of Sample.	CRUSHING STRENGTH—			REMARKS.
		Bed.	Height.		y. m. d.	lbs. oz.	Of Sample.	Per Square Inch.	Per Cubic Inch.	
PRISMS. 12" × 12" × 8"	a	12".03 × 12".14	8".09	Beds plast'r'd			lbs.			The prisms were at once tested combined in a pier, a next to driving head. With 560,000 lbs. load, began to flake off at joint a-b. Otherwise the pier gave scarcely any signs of yielding; it gave way suddenly under the ultimate load with a loud report. Prism a failed first, immediately followed by b; c last. A continuous seam first appeared along the three prisms, splitting off an entire corner of the pier; other and similar seams appeared rapidly in succession. The fragments of a were of pyramidal form with steep sides; the other two prisms were broken up by seams and cracks nearly parallel to the line of pressure
	b	11".98 × 12".08	8".08	"	1 10 28	259 4	654,800			
12" × 12" × 8"	c	12".08 × 12".10	8".08	"						

NOTE.—The cubes of neat cement from 8 inches upwards, except 8-inch cube a, and all the prisms, were each directly weighed; the weight of the other cement cubes was calculated.

GENERAL TABLE III.

Compressive Tests of Mortar made of Newark Company's Rosendale Cement.

Composition of Mortar: 1 Cement (dry measure), 3 Sand. Crushed between pine cushions.

Nominal Size of Cubes	Mark	Actual Size Bed	Height Sample	Height Incl'd'g Plaster	Size of Pine Cushions	Age when crushed (y. m.)	Weight of Sample (lbs. oz.)	Crushing Strength in Pounds Sample	Sq. Inch	Cubic Inch	Spread of Cushions parallel to Fibre	Indentation of Cushion Max.	Min.	Remarks.
2-inch	a	2"·00×2"·02	2"·07	2"·13	2½"×2½"×½"	1 10	— 9½	6,680	1,653	799	.06"&.05"	.09"	.04"	First crack at 6,600 lbs. Pyramidical fragments of cube.
"	b	2"·04×2"·03	1"·98	2"·07	2½"×2½"×½"	1 10	— 9½	5,330	1,206	609	.07"&.05"	.05"	.05"	First crack at 4,700 lbs. Pyramidical fragments of cube.
Av'ge									1,429	704				
4-inch	a	3"·98×4"·00	3"·98	4"·12	5"×5"×½"	1 10	4 2½	11,980	752	189	.16"&.08"	.07"	.02"	First crack at 11,600 lbs. Cube cleaved in lines parallel to grain of cushions.
"	b	3"·96×3"·96	4"·05	4"·14	5"×5"×½"	1 10	4 3½	12,100	765	189	.06"&.15"			First crack at 11,350 lbs. Cube cleaved in lines parallel to grain of cushions.
Av'ge									758	189				
6-inch	a	5"·93×6"·00	5"·95	6"·18	7"×7"×¾"	1 10	14 —	29,100	818	137	.00"&.16"	.04"	.01"	First crack at 28,050 lbs. Maximum indentation of cushion covered a surface 3¾"×5". Other portions of the cube were forced away from this section.
"	b	5"·93×5"·95	6"·00	6"·13	7"×7"×½"	1 10	14 —	27,600	782	130	.00"&.00"	.01"	.005"	First crack at 27,300 lbs. Pyramidical fragments.
Av'ge									800	133				
8-inch	a	8"·02×8"·06	7"·95	8"·08	9½"×9½"×½"	1 10	34 8	45,100	701	88	.14"&.27"	.18"	.01"	First crack at 45,100 lbs. Maximum indentation covered an area of 9¾"×8", adjacent portions of cube split off.
"	b	8"·10×8"·10	8"·12	8"·25	9½"×9½"×½"	1 10	34 12	46,800	713	88	.00"&.02"	.02"	.01"	First crack at 44,300 lbs. Pyramidical fragments.
Av'ge									707	88				

Note.—The specific weight of the mortar averaged 1.834, equivalent to a weight of 115.68 pounds per cubic foot.
The 2-inch cubes had the greatest density; their specific weight varied from 1.938 to 2.029.
In five of the eight sets of mortar cubes the strongest piece was also the heaviest one, per cubic foot; but in the other three sets the

GENERAL TABLE III.—(Continued.)

COMPRESSIVE TESTS OF MORTAR MADE OF NEWARK COMPANY'S ROSENDALE CEMENT.

Composition of Mortar: 1 Cement (dry measure), 3 Sand. Crushed between pine cushions.

Nominal Size of Cubes	Mark	Actual Size			Size of Pine Cushions	Age when crushed	Weight of Sample	Crushing Strength in Pounds.			Spread of Cushions parallel to Fibre.	Indentation of Cushion.		Remarks.
		Bed.	Height Sample.	Incl'd'g Plaster.				Sample.	Sq. Inch.	Cubic Inch.		Max.	Min.	
10-inch	a	10".00×10".00	10".10	10".15	11¼"×11¼"×¾"	y. m. 1 10	lbs. oz. 68 8	83,600	828	83	.07"&.05"	.07"	.07"	First crack at 81,600 lbs. Pyramidical fragments; cleavage parallel to grain of cushion.
" "	*b	10".10×10".06	10".04	10".15	11¼"×11¼"×¾"	1 10	69 8	108,000	1,063	105	.32"&.55"	.22"	.15"	First crack at 71,700 lbs. Cleavage on each compressed surface parallel to grain of cushion.
Av'ge.									943	94				
12-inch	a	12".00×12".04	12".00	12".08	14"×14"×1"	1 10	114 8	101,000	699	58	.04"&.04"	.03"	.02"	First crack at 100,200 lbs. Pyramidical fragments.
" "	*b	12".00×12".05	11".90	12".05	14"×14"×1"	1 10	114 8	97,100	671	56	.01"&.01"	.03"	.015"	First crack at 91,400 lbs. Pyramidical fragments; cleavage parallel to grain of cushion.
Av'ge.									635	57				
14-inch	a	14".00×13".98	13".90	14".03	16"×16"×1"	1 10	183	136,500	697	50	.02"&.03"	.06"	.02"	First crack at 132,000 lbs. Sides split off, leaving a solid core.
" "	*b	14".10×14".08	14".02	14".12	16"×16"×1"	1 10	188	145,600	733	52	.10"&.60"	.17"	.04"	First crack at 130,700 lbs. One cushion uniformly indented .04 inch, the other cushion unevenly indented from .04 to .17 inch.
Av'ge.									715	51				
16-inch	a	16".00×16".00	16".00	16".07	18"×18"×1"	1 10	271	157,000	613	38	.00"&.04"	.02"	.005"	First crack at 140,000 lbs. Sides split off, leaving a core of softer material.
" "	*b	16".04×16".02	16".04	16".17	18"×18"×1"	1 10	271	158,000	611	38	.00"&.03"			First crack at 147,000 lbs. Sides split off, leaving a core of softer material.
Av'ge.									612	38				

case was reversed. The two 10-inch cubes, which show a greater strength per square inch of compressed surface than the cubes of the preceding sets excepting the 2-inch cubes, had a greater specific weight (1.875 and 1.884 respectively) than any of the other cubes, with the exception already noted.

The pieces marked * were crushed with the fibres of the two cushions placed crosswise to each other, while the pieces not thus marked had the fibres of their cushions parallel. It was not noticed, however, that either one arrangement had any influence on the results, as compared with the other.

GENERAL TABLE III.—(*Continued.*)

COMPRESSIVE TESTS OF CONCRETE MADE OF NEWARK CO'S ROSENDALE CEMENT.

Composition of Concrete : 1 Cement, 3 Sand, 2 Gravel, 4 Broken Stone.

All bed-faces plastered. Some cubes crushed between pine cushions, and others without cushions.

Nominal Size of Cubes.	Mark.	Actual Size. Bed.	Height. Sample.	Includ'g Plaster.	How broken.	Size of Pine Cushions.	Age when crushed. y. m. d.	Weight of Sample. lbs. oz.	Crushing Strength in Pounds. Sample.	Sq. Inch.	Cubic Inch.	Spread of Cushion parallel to Grain.	Indentation of Cushion. Max.	Min.	Remarks.
4-inch	a	4".08 × 4".05	4".07	4".13	With Cushions.	5" × 5" × ½"	1 10 2	4 15	17,750	1,074	264	.07" & .03"	.13"	.01'	First crack at 15,000 lbs. Cube failed in detail, crushing from one side, cleaving in lines parallel to grain.
" "	b	4".06 × 4".07	4".02	4".12	With Cushions.	5" × 5" × ½"	1 10 2	4 14	16,380	991	247	.18" & .25"	.15"	.02"	First crack at 14,000 lbs. Otherwise like cube a.
6-inch	a	5".95 × 6".08	6".02	6".04	With Cushions.	7" × 7" × ½"	1 10 2	16 1	37,100	1,025	170	.14" & .16"	.06"	.005"	First crack at 35,000 lbs. Pyramidical fragments.
" "	b	6".02 × 6".09	6".12	6".19	Directly.	....	1 10 2	16 8	45,100	1,239	201				First crack at 41,500 lbs. Pyramidical fragments.
8-inch	a	8".07 × 8".09	8".00	8".14	With Cushions.	9½" × 9½" × ¾"	1 10 2	40 12	55,200	876	109.5	.25" & .08"	.30"	.05"	First crack at 43,600 lbs. Cube crushed in detail. Maximum indentation on a surface 5½" × 8".
" "	b	8".02 × 8".05	8".02	8".12	Directly.	....	1 10 2	39 —	77,100	1,194	149				First crack at 65,500 lbs. Pyramidical fragments.
10 inch	a	10".11 × 10".11	10".13	?	With Cushions.	11¼" × 11¼" × ¾"	1 10 2	79 8	117,700	1,151.5	113.7	.11" & .12"	.32"	.28"	First crack at 84,400 lbs. Pyramidical fragments.
" "	b	10".11 × 10".04	10".16	10".22	Directly.	....	1 10 2	78 —	120,000	1,182.2	116.4				First crack at 120,000 lbs. Pyramidical fragments.

GENERAL TABLE III.—(Concluded.)

COMPRESSIVE TESTS OF CONCRETE MADE OF NEWARK CO.'S ROSENDALE CEMENT.

Composition of Concrete: 1 Cement, 3 Sand, 2 Gravel, 4 Broken Stone.

All bed-faces plastered. Some cubes crushed between pine cushions, and others without cushions.

Nominal Size of Cubes.	Mark	Actual Size. Bed.	Height. Sample.	Height. Including Plaster.	How broken.	Size of Pine Cushions.	Age when crushed. y. m. d.	Weight of Sample. lbs. oz.	Crushing Strength in Pounds. Sample.	Sq. Inch.	Cubic Inch.	Spread of Cushion parallel to Grain.	Indentation of Cushion. Max.	Min.	Remarks.
12-inch	a	12".04 × 12".13	12".05	12".13	With Cushions.	14" × 14" × 1"	1 10 2	135 —	121,900	830.6	68.9	.06" & .08"	.15"	.05"	First crack at 108,-000 lbs. Compressed faces showed slight cleavage parallel to grain of cushions.
" "	b	12".06 × 12".04	12".00	12".02	Directly.		1 10 2	136 —	161,600	1,113	92.7				First crack at 150,-000 lbs. Pyramidical fragments.
14-inch	a	13".90 × 14".18	14".14	14".22	With Cushions.	16" × 16" × 1"	1 10 2	215 —	137,500	697.6	49.3	.08" & .30"	.20"	.01"	First crack at 134,-000 lbs. Compressed faces showed slight cleavage parallel to grain of cushions.
" "	b	14".09 × 14".05	14".04	14".13	Directly.		1 10 3	211 —	148,000	747.6	53.2				First crack at 147,-000 lbs. Pyramidical fragments.
16-inch	a	16".08 × 16".16	16".03	16".16	With Cushions.	18" × 18" × 1"	1 10 3	323 —	175,200	674.2	42.1	.42" & .08"	.12"	.005"	First crack at 130,-000 lbs. Compressed surfaces showed slight cleavage parallel to grain of cushions.
" "	b	16".03 × 16".10	16".04	16".16	Directly.		1 10 4	325 8	268,400	1,038.7	64.7				First crack at 250,-000 lbs. Pyramidical fragments.
18-inch	a	17".95 × 18".02	18".00	18".08	With Cushions.	20" × 20" × 1"	1 10 3	459 —	271,300	838.7	46.6	.5" & .6"	.25"	.02"	First crack at 220,-000 lbs. Compressed faces showed slight cleavage parallel to grain of cushions.
" "	b	18".06 × 17".62	18".00	18".19	Directly.		1 10 4	455 —	331,000	1,043.6	58				No preliminary crack. Pyramidical fragments.

GENERAL TABLE IVa.

COMPRESSIVE TESTS OF CUBES OF NORTON'S CEMENT.

Compressed surfaces coated with plaster. No pine cushions used.

MORTAR A.—Composition : 1 Cement Paste, 1¼ Sand.

NOMINAL SIZE.	Mark	ACTUAL SIZE.			Age when broken.	Weight of Sample.	CRUSHING STRENGTH IN POUNDS.			REMARKS.
		Bed.	Height.							
			Of Sample.	Including Plaster.	y. m. d.	lbs. oz.	Of Sample.	Per Square Inch.	Per Cubic Inch.	
4-inch Cube.	a	4″.06 × 4″.00	3″.98	4″.04	3 10 14	4 9½	33,000	2,032	510.6	First crack at 26,400 lbs.
4 ″	b	3″.88 × 4″.03	3″.99	4″.07	3 10 14	4 6¾	32,100	2,053	514.5	No preliminary cracking.
Average.....								2,042	512.5	
6-inch Cube.	a	6″.02 × 5″.99	6″.00	6″.12	3 10 14	14 12¾	49,700	1,378	229.7	No preliminary cracking.
6 ″	b	5″.98 × 6″.02	6″.03	6″.20	3 10 14	14 13	46,900	1,303	216.0	No preliminary cracking.
Average.....								1,340	222.8	
8-inch Cube.	a	8″.05 × 8″.03	8″.11	8″.14	3 10 14	37 12	106,000	1,640	202.2	Cracks appeared at time of.
8 ″	b	8″.05 × 8″.05	7″.99	8″.00	3 10 14	36 12	120,000	1,852	231.7	Cracks appeared at 118,000 lbs. Pyramidical fragments.
Average.....								1,746	216.9	
12-inch Cube.	a	12″.03 × 12″.07	12″.03	12″.24	3 10 14	118 8	192,500	1,326	110.2	Cracks in sight at 196,100 lbs.
12 ″	b	12″.02 × 12″.02	12″.07	12″.11	3 10 14	118 12	197,400	1,386	113.2	
Average.....								1,356	111.7	
16-inch Cube.	a	16″.00 × 16″.02	16″.05	16″.13	3 10 14	284 —	321,200	1,254	78.1	No preliminary cracks.
16 ″	b	16″.05 × 16″.08	16″.08	16″.17	3 10 14	284 8	320,000	1,249	77.4	No preliminary cracks.
Average.....								1,247	77.6	

GENERAL TABLE IVA.—(Concluded.)

CONCRETE A.—Composition: 1 Cement Paste, 1½ Sand, 6 Broken Stone.

Nominal Size	Mark	Actual Size — Bed	Actual Size — Height, Of Sample	Actual Size — Height, Including Plaster	Age when broken (y. m. d.)	Weight of Sample (lbs. oz.)	Crushing Strength — Of Sample	Crushing Strength — Per Square Inch	Crushing Strength — Per Cubic Inch	Remarks
4 inch Cube	a	4".20 × 4".03	4".10	4".13	3 10 14	5 5¾	37,950	2,320	567.1	No preliminary crack.
4 " "	b	4".05 × 4".05	4".11	4".12	3 10 14	5 5⅜	38,100	2,323	565.2	No preliminary crack.
Average								2,322	566.2	
6-inch Cube	a	6".07 × 6".10	6".03	6".04	3 10 14	17 4¼	33,100	908.9	150.7	First crack in sight at 33,000 lbs.
6 " "	b	6".09 × 6".07	6".00	6".05	3 10 14	17 9	37,200	1,016.3	169.4	First crack in sight at 35,200 lbs.
Average								962.6	160.0	
8-inch Cube	a	8".03 × 8".07	8".06	8".16	3 10 14	43 8	87,600	1,352	165.6	No preliminary signs of yielding.
8 " "	b	8".05 × 8".04	8".04	8".07	3 10 15	43 —	97,900	1,516	189.0	No preliminary signs of yielding.
Average								1,434	177.3	
12-inch Cube	a	12".09 × 12".00	12".02	12".12	3 10 14	148 8	218,000	1,503	125	Cracks in sight at 184,000 lbs.
12 " "	b	12".00 × 12".00	12".05	12".15	3 10 14	148 8	232,900	1,617	134	Cracks in sight at 200,000 lbs.
Average								1,560	129.5	
16-inch Cube	a	16".10 × 16".07	16".05	16".20	3 10 15	353 —	379,200	1,466	91.3	Snapping sounds at 270,000 lbs., cracks in sight at 360,000 lbs.
16 " "	b	16".04 × 16".05	16".10	16".27	3 10 14	352 8	368,000	1,429	88.8	Cracks in sight at 360,000 lbs.
Average								1,447.5	90.0	

GENERAL TABLE IVB.

COMPRESSIVE TESTS OF CUBES OF NORTON'S CEMENT.

Compressed surfaces coated with plaster.   No pine cushions used.

MORTAR B.—Composition : 1 Cement Paste, 3 Sand.

NOMINAL SIZE.	Mark.	ACTUAL SIZE. Bed.	Height. Of Sample	Height. Including Plaster.	Age when broken. y. m. d.	Weight of Sample. lbs. oz.	CRUSHING STRENGTH IN POUNDS. Of Sample.	Per Square Inch.	Per Cubic Inch.	REMARKS.
4-inch Cube. 4 "	a	3".98 × 3".94	3".90	3".92	3 10 9	4 4	23,250	1,483	380	
	b	3".98 × 4".00	4".00	4".06	3 10 9	4 7¾	18,560	1,165.8	291.5	
Average....								1,324.4	335.8	
6-inch Cube. 6 "	a	6".02 × 6".03	5".98	6".01	3 10 9	14 10	28,390	779.6	130.4	
	b	6".02 × 6".01	5".94	6".10	3 10 9	14 9½	26,050	721.2	121.4	
Average....								750.4	125.9	
8-inch Cube. 8 "	a	7".96 × 8".04	8".05	8".18	3 10 9	35 —	54,280	848.5	105.4	No preliminary signs of yielding.
	b	8".05 × 8".02	8".00	8".10	3 10 9	35 —	47,230	731.9	91.5	
Average....								790.2	98.5	
12-inch Cube. 12 "	a	12".02 × 12".06	12".00	12".08	3 10 9	116 —	98,500	679.5	56.5	
	b	12".11 × 12".07	12".11	12".14	3 10 10	116 8	101,600	693.7	57.4	
Average....								687.6	57.0	
16-inch Cube. 16 "	a	16".10 × 16".11	16".10	?	3 10 10	277 8	104,200	748.7	46.5	
	b	16".07 × 16".00	16".09	?	3 10 10	277 8	176,750	687.4	42.7	
Average....								778.1	44.6	

GENERAL TABLE IVB.—(Concluded.)

CONCRETE B.—Composition : 1 Cement Paste, 3 Sand, 6 Broken Stone.

Nominal Size.	Mark.	Actual Size. Bed.	Height. Of Sample.	Height. Including Plaster.	Age when broken. y. m. d.	Weight of Sample. lbs. oz.	Crushing Strength in Pounds. Of Sample.	Per Square Inch.	Per Cubic Inch.	Remarks.
4-inch Cube. 4 "	a	3".98 × 4".00	4".07	4".11	3 10 10	5 4	24,700	1,551.5	381.2	First cracks at 23,150 lbs.
	b	3".98 × 4".00	4".02	4".23	3 10 10	4 12	27,300	1,714.8	426.6	No preliminary signs of yielding.
Average...								1,633.2	403.9	
6-inch Cube. 6 "	a	6".02 × 6".00	5".90	6".07	3 10 10	17 4	36,450	1,009	171.0	No preliminary signs of yielding.
	b	5".95 × 6".00	6".00	6".07	3 10 10	17 5	35,380	991	165.2	First cracks at 35,100 lbs.
Average...								1,000	168.1	
8-inch Cube. 8 "	a	8".02 × 8".00	8".02	8".16	3 10 10	42 —	56,400	879	109.6	First cracks at 56,000 lbs.
	b	8".00 × 8".15	8".05	8".24	3 10 10	42 8	55,000	843.6	104.8	No preliminary signs of yielding.
Average...								861.3	107.2	
12-inch Cube. 12 "	a	12".01 × 12".11	12".03	12".17	3 10 10	140 —	112,650	774.5	64.3	No preliminary signs of yielding.
	b	12".06 × 12".05	12".05	12".14	3 10 10	140 —	109,900	756.2	62.7	No preliminary signs of yielding.
Average...								765.4	63.5	
16-inch Cube. 16 "	a	16".14 × 16".03	16".12	16".21	3 10	339 —	222,100	858.4	53.2	No preliminary signs of yielding. (See Special Table VII.)
	b	16".12 × 16".10	16".14	16".24	3 10	339 8	215,000	828.4	51.3	First cracks at 210,000 lbs.
Average...								843.4	57.3	

GENERAL TABLE V.

COMPRESSIVE TESTS OF CUBES OF NATIONAL PORTLAND CEMENT.

Compressed surfaces coated with plaster. No pine cushions used.

MORTAR C.—Composition: 1 vol. Cement Paste, 3 vols. Sand.

NOMINAL SIZE.	Mark.	ACTUAL SIZE. Bed.	Height. Of Sample.	Height. Including Plaster.	Age when broken. y. m. d.	Weight of Sample. lbs. oz.	CRUSHING STRENGTH IN POUNDS. Of Sample.	Per Square Inch.	Per Cubic Inch.	REMARKS.
4-inch Cube.	a	4″.03 × 4″.04	4″.00	4″.11	3 10 3	4 9	58,800	3,612	903	No preliminary signs of yielding.
4 "	b	4″.02 × 3″.98	4″.05	4″.10	3 10 3	4 8	57,600	3,288	812	
Average								3,450	857	
6-inch Cube.	a	6″.02 × 5″.99	5″.97	6″.09	3 10 3	14 11¼	99,800	2,768	464	
6 "	b	6″.00 × 5″.98	5″.93	5″.99	3 10 3	14 9¾	91,200	2,512	479	
Average								2,655	446	
8-inch Cube.	a	8″.08 × 8″.04	8″.01	8″.13	3 10 5	35 8	168,000	2,386	323	
8 "	b	8″.01 × 7″.96	8″.13	8″.25	3 10 5	36 —	150,000	2,353	289	
Average								2,469	306	
12-inch Cube.	a	12″.00 × 12″.06	12″.07	12″.15	3 10 5	125 8	357,400	2,472	205	
12 "	b	12″.02 × 12″.00	12″.10	12″.15	3 10 5	125 —	345,600	2,396	191	
Average								2,434	198	
16-inch Cube.	a	16″.12 × 16″.12	16″.22	16″.24	3 10 5	283 —	650,000	2,501	154	
16 "	b	16″.04 × 16″.08	16″.12	16″.20	3 10 5	283 —	654,500	2,537	157	
Average								2,519	155	

GENERAL TABLE V.—*(Concluded.)*

CONCRETE C.—Composition: 1 vol. Cement Paste, 3 vols. Sand, 6 vols. Broken Stone.

NOMINAL SIZE.	Mark.	ACTUAL SIZE. Bed.	Height. Of Sample.	Height. Including Plaster.	Age when broken. y. m. d.	Weight of Sample. lbs. oz.	CRUSHING STRENGTH IN POUNDS. Of Sample.	Per Square Inch.	Per Cubic Inch.	REMARKS.
4-inch Cube.	a	4″.04 × 4″.00	4″.02	4″.09	3 10 6	5 4	63,400	3,923	976	
4 "	b	3″.98 × 4″.03	4″.04	4″.08	3 10 6	5 5	65,850	4,105	1,016	
Average...							......	4,014	996	
6-inch Cube.	a	6″.01 × 6″.02	6″.00	6″.08	3 10 6	17 5	87,950	2,436	405	No preliminary signs of yielding.
6 "	b	5″.97 × 5″.97	5″.99	6″.02	3 10 6	16 12	109,600	2,823	478	
Average...							......	2,629	492	
8 inch Cube.	a	8″.04 × 7″.99	8″.11	8″.24	3 10 6	43 —	196,500	3,058	377	Cracks in sight at 365,500 lbs.
8 "	b	8″.05 × 8″.03	8″.18	8″.21	3 10 6	43 —	193,500	2,993	396	No preliminary signs of yielding.
Average...							......	3,025	371	
12-inch Cube.	a	12″.00 × 12″.00	12″.10	12″.19	3 10 6	143 8	367,000	2,540	210	
12 "	b	12″.00 × 12″.03	12″.10	12″.18	3 10 6	143 8	410,000	2,840	235	
Average...							......	2,690	222	
16-inch Cube.	a	16″.06 × 16″.15	16″.11	16″.19	3 10 6	345 —	747,000	2,880	179	Cracks in sight at 700,000 lbs. Broken only by repeated application of maximum load of 800,000 lbs., alternating with reduction of pressure to 5,000 lbs. (See Special Table IX.)
16 "	b	16″.17 × 16″.08	16″.16	16″.24	3 10 6	352 —	800,000+	3,077+	190+	
Average...							......	2,978+	185+	

GENERAL TABLE VI.

COMPRESSIVE TESTS OF BRICK PIERS.

Each pier was built in six courses of common, hard, North River brick, one and one half brick in cross-section. The mortar consisted of one part of Newark Company's Rosendale cement and two parts of sand. The mortar-joints were about ⅜ inch thick; the piers were built to represent ordinary brickwork.

Each pier was finished off at bottom and top with a smooth-faced slab of North River bluestone.

The age of the piers when crushed was 21½ months.

No.	Dimensions. Cross-section.	Height. Brick-work.	Height. Including Bluestone	Weight of— Brick-work only	Weight of— Entire Pier.	Crushing Strength in Pounds. Of Pier.	Crushing Strength in Pounds. Per Sq. Foot.	Remarks.
I.	12".00 × 12".00	......	22".42	154 lbs.	238 lbs.	291,000	291,000	First snapping sounds at 200,000 lbs. Longitudinal cracks appeared in two courses under a load of 240,000 lbs.
II.	11".90 × 12".00	16".53	22".08	151 "	233 "	260,000	262,185	Snapping sounds at 160,000 lbs., but no cracks visible. Cracks were developed in three courses under a load of 220,000 lbs.
III.	12".00 × 12".00	16".32	22".38	154 "	241 "	260,000	260,000	Snapping sounds at 140,000 lbs., but no cracks seen. When a load of 200,000 lbs. was reached, cracks were visible in three courses.
IV.	12".00 × 12".00	16".25	22".50	153 "	240 "	280,000	280,000	Snapping sounds at 190,000 lbs. At 200,000 lbs. cracks were in sight in two courses.
V.	12".00 × 12".00	15".97	23".22	148 "	251 "	250,000	250,000	Under a load of 200,000 lbs. the third course began to flake off. With a load of 200,000 lbs. a general development of longitudinal cracks set in.
VI.	11".75 × 12".00	15".88	21".98	147 "	230 "	251,000	256,340	First cracks appeared under a load of 150,000 lbs. in the second course.
Average..........							266,587 pounds, or 115 gross tons.	

SPECIAL TABLE I.

SHOWING AMOUNT OF COMPRESSION AND SET OF CUBES OF HAVERSTRAW FREESTONE (N. Y.).

8-INCH FREESTONE CUBE, MARKED *a* ; BEDS PLASTERED.

Actual size: Bed = 7″.99 × 7″.99; Height = 7″.99 (or 8″.15 including plaster); Weight, 39 pounds.

LOAD.	INCH.		LOAD.	INCH.		LOAD.	INCH.	
Pounds.	Compression.	Set.	Pounds.	Compression.	Set.	Pounds.	Compression.	Set.
5,000	....	....	100,000	.0205	....	200,000	.0310	....
10,000	.0021	....	5,000	....	.0078	5,000	....	.0105
20,000	.0062	....	100,000	.0210	....	200,000	.0315	....
30,000	.0093	....	110,000	.0220	....	220,000	.0332	....
40,000	.0112	....	120,000	.0230	....	240,000	.0355	....
50,000	.0132	....	130,000	.0240	....	260,000	.0380	....
5,000	... ...	.0055	140,000	.0250	....	280,000	.0410	....
50,000	.0138	....	150,000	.0260	....	300,000	.0442	....
60,000	.0150	....	5,000	....	.0090	310,000	.0460	... ...
70,000	.0165	....	150,000	.0260	....	320,000	.0480	....
80,000	.0179	....	160,000	.0270	....	330,000	.0495	......
90,000	.0192	... .	180,000	.0290	... .	337,000	broken	....

8-INCH FREESTONE CUBE, MARKED *b* ; BEDS PLASTERED.

Actual size: Bed = 8″.05 × 8″.16; Height = 8″.00 (or 8″.14 including plaster); Weight, 41¼ pounds.

LOAD.	INCH.		LOAD.	INCH.		LOAD.	INCH.	
Pounds.	Compression.	Set.	Pounds.	Compression.	Set.	Pounds.	Compression.	Set.
5,000	....	....	140,000	.0225	....	300,000	.0408	....
10,000	.0015	....	160,000	.0250	... .	310,000	.0420	....
20,000	.0045	....	180,000	.0268	....	320,000	.0435	....
40,000	.0092	....	200,000	.0287	....	330,000	.0450	....
60,000	.0103	....	5,000	....	.0110	340,000	.0465	... ...
80,000	.0155	....	200,000	.0290	....	350,000	.0480	....
100,000	.0180	....	220,000	.0305	....	360,000	.0500	....
5,000	....	.0078	240,000	.0330	... .	370,000	.0512	....
100,000	.0182	....	260,000	.0355	....	438,400	broken	... ...
120,000	.0205	....	280,000	.0380	....	....	....	....

SPECIAL TABLE I.—(*Continued.*)

8-Inch Freestone Cube, marked c; Beds Plastered.

Actual size: Bed=8".00 × 8".03; Height = 8".00 (or 8".07 including plaster); Weight, 39¾ pounds.

Load. Pounds.	Inch. Compression.	Set.	Load. Pounds.	Inch. Compression.	Set.	Load. Pounds.	Inch. Compression.	Set.
5,000	....	....	120,000	.0190	....	260,000	.0335	....
10,000	.0022	....	140,000	.0210	....	280,000	.0360	....
20,000	.0050	....	160,000	.0230	....	300,000	.0380	....
40,000	.0070	....	180,000	.0250	....	320,000	.0402	....
60,000	*.0120	....	200,000	.0270	....	340,000	.0425	....
80,000	.0148	....	5,000	....	.0098	360,000	.0450	....
100,000	.0165	....	200,000	.0275	....	370,000	.0472	....
5,000	....	.0065	220,000	.0295	....	380,000	.0488	....
100,000	.0170	....	240,000	.0315	....	388,000	.0515	broken

8-Inch Freestone Cube, marked d; Beds Plastered.

Actual size: Bed = 8".02 × 8".02; Height=7".96 (or 8".04 including plaster); Weight, 39½ pounds.

Load. Pounds.	Inch. Compression.	Set.	Load. Pounds.	Inch. Compression.	Set.	Load. Pounds.	Inch. Compression.	Set.
5,000	....	....	140,000	.0215	....	280,000	.0372	....
10,000	.0012	....	160,000	.0240	....	300,000	.0395	....
20,000	.0042	....	180,000	.0260	....	320,000	.0425	....
40,000	.0082	....	200,000	.0280	....	340,000	.0450	....
60,000	.0115	....	5,000	....	.0110	350,000	.0462	....
80,000	.0145	....	200,000	.0280	....	360,000	.0480	....
100,000	.0170	....	220,000	.0300	....	370,000	.0495	....
5,000	....	.0070	240,000	.0325	....	380,000	.0510	sudden yielding
100,000	.0172	....	260,000	.0348	....	387,000	.0530	
120,000	.0190	....	270,000	.0360	....	395,700	broken	

9-Inch Freestone Cube, marked a; Beds Plastered.

Actual size: Bed = 9".07 × 8".99; Height = 8".96 (or 9".05 including plaster); Weight, 56 pounds.

Load. Pounds.	Inch. Compression.	Set.	Load. Pounds.	Inch. Compression.	Set.	Load. Pounds.	Inch. Compression.	Set.
5,000	....	....	180,000	.0375	....	5,000	....	.0270
40,000	.0095	....	200,000	.0400	....	300,000	.0542	....
80,000	.0172	....	5,000	....	.0220	340,000	.0582	....
100,000	.0220	....	260,000	.0410	....	380,000	.0625	....
5,000	....	.0110	240,000	.0460	....	400,000	.0642	....
100,000	.0222	....	280,000	.0510	....	470,400	broken	....
140,000	.0330	....	300,000	.0532	....			

SPECIAL TABLE I.—(*Continued.*)

9-INCH FREESTONE CUBE, MARKED *b*; BEDS PLASTERED.

Actual size: Bed = 9".03 × 9".00; Height=8".97 (or 9".05 including plaster); Weight, 57¾ pounds.

| Load. | Inch. | | Load. | Inch. | | Load. | Inch. | |
Pounds.	Compression.	Set.	Pounds.	Compression.	Set.	Pounds.	Compression.	Set.
5,000	....	....	5,000	....	.0100	5,000	....	.0160
20,000	.0045	....	200,000	.0265	....	400,000	.0475	....
40,000	.0080	....	240,000	.0300	....	420,000	.0490	....
80,000	.0138	....	280,000	.0338	....	440,000	.0510	....
100,000	.0160	....	300,000	.0360	....	460,000	.0530	....
5,000	....	.0070	5,000	....	.0130	480,000	.0552	....
100,000	.0162	....	300,000	.0365	....	490,000	.0562	....
140,000	.0200	....	340,000	.0400	....	536,000	.0577	....
180,000	.0240	....	380,000	.0440	....	568,000	broken	....
200,000	.0262	....	400,000	.0460	....			

9-INCH FREESTONE CUBE, MARKED *c*; BEDS PLASTERED.

Actual size: Bed = 9".02 × 9".04; Height=9".01 (or 9".05 including plaster); Weight, 57¾ pounds.

| Load. | Inch. | | Load. | Inch. | | Load. | Inch. | |
Pounds.	Compression.	Set.	Pounds.	Compression.	Set.	Pounds.	Compression.	Set.
5,000	....	....	200,000	.0285	....	420,000	.0500	....
20,000	.0052	....	300,000	.0380	....	440,000	.0515	....
40,000	.0088	....	5,000	....	.0150	460,000	.0530	....
80,000	.0150	....	300,000	.0385	....	480,000	.0548	....
100,000	.0175	....	340,000	.0420	....	500,000	.0560	....
5,000	....	.0090	380,000	.0460	....	520,000	.0580	....
100,000	.0178	....	400,000	.0475	....	540,000	.0592	....
200,000	.0280	....	5,000	....	.0182	550,000	.0605	....
5,000	....	.0122	400,000	.0488	....	643,000	broken	....

9-INCH FREESTONE CUBE, MARKED *d*; BEDS PLASTERED.

Actual size: Bed = 8".99 × 9".04; Height= 8".92 (or 8".99 including plaster); Weight, 56½ pounds.

| Load. | Inch. | | Load. | Inch. | | Load. | Inch. | |
Pounds.	Compression.	Set.	Pounds.	Compression.	Set.	Pounds.	Compression.	Set.
5,000	....	....	5,000	....	.0165	380,000	.0567	....
20,000	.0050	....	200,000	.0350	....	400,000	.0588	....
40,000	.0100	....	300,000	.0472	....	5,000	....	.0252
80,000	.0170	....	5,000	....	.0218	400,000	.0595	....
100,000	.0200	....	300,000	.0480	....	410,000	.0610	....
5,000	....	.0099	320,000	.0500	....	420,000	.0615	....
100,000	.0205	....	340,000	.0520	....	440,000	.0635	....
200,000	.0345	....	360,000	.0540	....	445,000	.0650	broken

SPECIAL TABLE I.—(*Continued.*)

10-Inch Freestone Cube, marked a: Beds Plastered.

Actual size: Bed = 10".02 × 9".96; Height = 10".01 (or 10".07 including plaster); Weight, 79¾ pounds.

Load. Pounds.	Inch. Compression.	Set.	Load. Pounds.	Inch. Compression.	Set.	Load. Pounds.	Inch. Compression.	Set.
5,000	....	....	200,000	.0390	....	5,000	....	.0300
20,000	.0055	....	5,000	....	.0230	400,000	.0610	....
40,000	.0100	....	200,000	.0400	....	440,000	.0635	....
80,000	.0180	....	300,000	.0510	....	480,000	.0665	....
100,000	.0220	....	5,000	....	.0275	500,000	.0685	cracked
5,000	....	.0130	300,000	.0520	....	520,000	broken	
100,000	.0222	....	400,000	.0600	....			

10-Inch Freestone Cube, marked b; Beds Plastered.

Actual size: Bed = 10".00 × 9".80; Height = 10".01 (or 10".12 including plaster); Weight, 77½ pounds.

Load. Pounds.	Inch. Compression.	Set.	Load. Pounds.	Inch. Compression.	Set.	Load. Pounds.	Inch. Compression.	Set.
5,000	....	....	300,000	.0305	....	500,000	.0485	....
20,000	.0038	....	5,000	....	.0100	540,000	.0520	....
40,000	.0070	....	300,000	.0310	....	580,000	.0560	....
80,000	.0120	....	400,000	.0382	....	600,000	.0570	....
100,000	.0145	....	5,000	....	.0117	5,000	....	.0178
5,000	....	.0062	400,000	.0390	....	600,000	.0592	....
100,000	.0148	....	440,000	.0420	....	620,000	.0615	....
200,000	.0230	....	480,000	.0457	....	640,000	.0632	....
5,000	....	.0080	500,000	.0478	....	650,500	failed suddenly,	
200,000	.0232	....	5,000	....	.0142		without warning	

10-Inch Freestone Cube, marked c; Beds Plastered.

Actual size: Bed = 10".00 × 9".96; Height = 10".01 (thickness of plaster not noted); Weight 78¼ pounds.

Load. Pounds.	Inch. Compression.	Set.	Load. Pounds.	Inch. Compression.	Set.	Load. Pounds.	Inch. Compression.	Set.
5,000	....	....	5,000	....	.0071	300,000	.0475	....
20,000	.0030	....	200,000	.0225	....	5,000	....	....
40,000	.0062	....	300,000	.0300	....	Not broken under maximum		
80,000	.0115	....	5,000	....	.0090	load of 800,000 pounds.		
100,000	.0132	....	300,000	.0305	....			
5,000	....	.0049	400,000	.0390	....			
100,000	.0137	....	5,000	....	.0215			
200,000	.0220	....	400,000	.0390	....			

SPECIAL TABLE I.—(Continued.)

10-INCH FREESTONE CUBE, MARKED d: BEDS PLASTERED.

Actual size: Bed = 10".00 × 9".98; Height = 10".00 (or 10".09 including plaster); Weight, 78¼ pounds.

LOAD. Pounds.	INCH. Compression.	Set.	LOAD. Pounds.	INCH. Compression.	Set.	LOAD. Pounds.	INCH. Compression.	Set.
5,000	....	....	200,000	.0250	....	5,000	....	.0132
20,000	.0040	....	5,000	....	.0085	400,000	.0418	....
40,000	.0078	....	200,000	.0250	....	500,000	.0520	....
80,000	.0135	....	300,000	.0325	....	5,000	....	.0170
100,000	.0157	....	5,000	....	.0110	644,000	broken	....
5,000	....	.0062	300,000	.0329	....	....	....	....
100,000	.0160	....	400,000	.0412	....	....	....	....

11-INCH FREESTONE CUBE, MARKED a; BEDS PLASTERED.

Actual size: Bed = 11".05 × 11".00; Height = 10".92 (or 11".09 including plaster); Weight, 105 pounds.

LOAD. Pounds.	INCH. Compression.	Set.	LOAD. Pounds.	INCH. Compression.	Set.	LOAD. Pounds.	INCH. Compression.	Set.
5,000	....	....	200,000	.0262	....	5,000	....	.0193
40,000	.0072	....	300,000	.0340	....	500,000	.0492	....
80,000	.0125	....	5,000	....	.0152	600,000	.0562	....
100,000	.0152	....	300,000	.0350	....	5,000	....	.0220
5,000	....	.0075	400,000	0412	....	600,000	.0575	....
100,000	.0154	....	5,000	....	.0175	770,000	cracked	....
200,000	.0260	....	400,000	.0417	....	791,000	broken	....
5,000	....	.0120	500,000	.0485	....	....	....	....

11-INCH FREESTONE CUBE, MARKED b; BEDS PLASTERED.

Actual size: Bed = 11".10 × 10".96; Height = 11".01 (or 11".08 including plaster); Weight, 106½ pounds.

LOAD. Pounds.	INCH. Compression.	Set.	LOAD. Pounds.	INCH. Compression.	Set.	LOAD. Pounds.	INCH. Compression.	Set.
5,000	....	....	200,000	.0242	....	5,000	....	.0140
40,000	.0072	....	300,000	0308	....	500,000	.0455	....
80,000	.0122	....	5,000	....	.0105	600,000	.0530	....
100,000	.0145	....	300,000	.0312	....	5,000	....	.0155
5,000	....	.0060	400,000	.0380	....	600,000	.0540	....
100,000	.0150	....	5,000	....	.0120	770,000	cracked	....
200,000	.0240	....	400,000	.0380	....	785,000	broken	....
5,000	....	.0082	500,000	.0450	....	....	....	....

SPECIAL TABLE I.—(*Continued.*)

11-INCH FREESTONE CUBE, MARKED c; BEDS PLASTERED.

Actual size: Bed = 11".00 x 11".00; Height = 10".97 (or 11".01 including plaster); Weight, 104½ pounds.

Load.	Inch.		Load.	Inch.		Load.	Inch.	
Pounds.	Compression.	Set.	Pounds.	Compression.	Set.	Pounds.	Compression.	Set.
5,000	....	....	200,000	.0272	....	5,000	....	.0178
40,000	.0081	....	300,000	.0350	....	500,000	.0507	....
80,000	.0145	....	5,000	.. ..	.0140	600,000	.0575	....
100,000	.0170	....	300,000	.0350	....	5,000	....	.0210
5,000	....	.0080	400,000	.0420	....	500,000	.0580	
100,000	.0175	....	5,000	....	.0160	778,000	cracked	....
200,000	.0270	... .	400,000	.0425	....	779,000	broken	....
5,000	....	.0118	500,000	.0500	....	... .	....	....

11-INCH FREESTONE CUBE, MARKED d; BEDS PLASTERED.

Actual size: Bed = 11".10 x 11".05; Height = 11".02 (or 11".16 including plaster); Weight, 106½ pounds.

Load.	Inch.		Load.	Inch.		Load.	Inch.	
Pounds.	Compression.	Set.	Pounds.	Compression.	Set.	Pounds.	Compression.	Set.
5,000	....	....	200,000	.0230	....	5,000	... .	.0180
40,000	.0065	. ...	300,000	.0300	....	500,000	.0510	
80,000	.0120	....	5,000	....	.0099	600,000	.0600	....
100,000	.0140	....	300,000	.0310	....	5,000	....	.0220
5,000	....	.0052	400,000	.0388	....	600,000	.0615	....
100,000	.0140	....	5,000	....	.0132	769,000	broken	....
200,000	.0228	....	400,000	.0392	....	....	....	
5,000	....	.0078	500,000	.0500	....	....	....	... .

12-INCH FREESTONE CUBE, MARKED a; BEDS PLASTERED.

Actual size: Bed = 12".00 x 11".95; Height = 12".01 (or 12".05 including plaster); Weight, 139½ pounds.

Load.	Inch.		Load.	Inch.		Load.	Inch.	
Pounds.	Compression.	Set.	Pounds.	Compression.	Set.	Pounds.	Compression.	Set.
5,000	....	....	300,000	.0355	....	600,000	.0555	....
40,000	.0095	....	5,000	....	.0135	5,000	....	.0188
80,000	.0160	....	300,000	.0360	....	600,000	.0560	....
100,000	.0185	....	400,000	.0420	....	700,000	.0620	....
5,000	....	.0085	5,000	....	.0150	5,000	....	.0202
100,000	.0192	....	400,000	.0425	....	700,000	.0632	....
200,000	.0282	....	500,000	.0487	....	800,000	.0690	....
5,000	....	.0115	5,000	....	.0170	5,000	....	.0225
200,000	.0288	....	500,000	.0492	... .	Cube removed from the press.		

SPECIAL TABLE I.—(Continued.)

12-INCH FREESTONE CUBE, MARKED *b*; BEDS PLASTERED.

Actual size: Bed = 12″.00 × 12″.00; Height = 12″.04 (or 12″.23 including plaster); Weight, 138 pounds.

Load.	Inch.		Load.	Inch.		Load.	Inch.	
Pounds.	Compression.	Set.	Pounds.	Compression.	Set.	Pounds.	Compression.	Set.
5,000	....	....	300,000	.0265	....	600,000	.0430	....
40,000	.0060	....	5,000	....	.0082	5,000	....	.0128
80,000	.0110	....	300,000	.0270	....	600,000	.0440	....
100,000	.0130	....	400,000	.0320	....	700,000	.0500	....
5,000	....	.0050	5,000	....	.0098	5,000	....	.0150
100,000	.0130	....	400,000	.0320	....	700,000	.0510	....
200,000	.0205	....	500,000	.0370	....	800,000	.0585	....
5,000	....	.0070	5,000	....	.0110	5,000	....	.0180
200,000	.0210	....	500,000	.0370	....	5,000	reduced to .0172 after 1 hour's rest. Cube removed from the press.	

12-INCH FREESTONE CUBE, MARKED *c*; BEDS PLASTERED.

Actual size: Bed = 11″.96 × 12″.00; Height = 12″.00 (or 12″.20 including plaster); Weight, 135½ pounds.

Load.	Inch.		Load.	Inch.		Load.	Inch.	
Pounds.	Compression.	Set.	Pounds.	Compression.	Set.	Pounds.	Compression.	Set.
5,000	....	....	300,000	.0355	....	600,000	.0560	....
40,000	.0102	....	5,000	....	.0142	5,000	....	.0200
80,000	.0170	....	300,000	.0355	....	600,000	.0570	....
100,000	.0192	....	400,000	.0420	....	700,000	.0658	... .
5,000	....	.0090	5,000	....	.0160	5,000	....	.0225
100,000	.0200	....	400,000	.0420	....	700,000	.0675	....
200,000	.0288	....	500,000	.0485	....	740,000	.0727	cracked
5,000	....	.0120	5,000	....	.0180	764,000	broken	....
200,000	.0290	....	500,000	.0490	....	....	....	....

12-INCH FREESTONE CUBE, MARKED *d*; BEDS PLASTERED.

Actual size: Bed = 11″.96 × 11″.90; Height = 12″.01 (or 12″.14 including plaster); Weight, 135¾ pounds.

Load.	Inch.		Load.	Inch.		Load.	Inch.	
Pounds.	Compression.	Set.	Pounds.	Compression.	Set.	Pounds.	Compression.	Set.
5,000	....	....	300,000	.0248	....	600,000	.0420	....
40,000	.0050	....	5,000	....	.0065	5,000	....	.0098
80,000	.0090	....	300,000	.0250	....	600,000	.0425	....
100,000	.0110	....	400,000	.0300	....	700,000	.0495	....
5,000	....	.0035	5,000	....	.0078	5,000	....	.0115
100,000	.0112	....	400,000	.0305	....	700,000	.0500	....
200,000	.0185	....	500,000	.0355	....	800,000	.0565	....
5,000	....	.0050	5,000	....	.0085	5,000	....	.0140
200,000	.0188	....	500,000	.0360	....	....	Cube removed from the press.	

SPECIAL TABLE I.—(*Concluded.*)

PIER OF CUBES OF HAVERSTRAW FREESTONE; DRY JOINTS

THREE 12-INCH CUBES, MARKED *a*, *b*, AND *d*, RESPECTIVELY ; BEDS PLASTERED.

Each of these cubes had been previously tested up to the maximum load of 800,000 pounds without breaking it.

Actual size: Cube *a*—Bed = 12".00 × 11".95; Height = 12".01 (or 12".05 including plaster) ;
Weight, 139½ pounds.

Cube *b*—Bed = 12".00 × 12".00; Height = 12".04 (or 12".23 including plaster) ;
Weight, 138 pounds.

Cube *d*—Bed = 11".96 × 11".90; Height = 12".01 (or 12".14 including plaster) ;
Weight, 135¾ pounds.

LOAD.	INCH.		LOAD.	INCH.		LOAD.	INCH.	
Pounds.	Compression.	Set.	Pounds.	Compression.	Set.	Pounds.	Compression.	Set.
5,000	....	....	300,000	.0805		5,000	....	.0112
40,000	.0210	....	5,000	....	.0060	600,000	.1220	....
80,000	.0350	....	300,000	.0805		700,000	.1370	... ...
100,000	.0415	....	400,000	.0942	....	5,000	....	.0150
5,000	....	.0025	5,000	....	.0080	700,000	.1400 { crack at	{ *a* in sight,
100,000	.0422	.. ..	500,000	.1075	....	748,000	failed suddenly	
200,000	.0638	....	5,000	....	.0095		with loud report.	
5,000	....	.0042	500,000	.1080	....	.... ..	....	....
200,000	.0638	....	600,000	.1210	....	....	....	....

SPECIAL TABLE II.

SHOWING AMOUNT OF COMPRESSION AND SET OF SPECIMENS OF NEAT PORTLAND (DYCKERHOFF) CEMENT.

8-INCH CUBE, MARKED *Db* ; BEDS NOT PLASTERED.

Actual size: Bed = 8".01 × 8".03; Height = 7".99; Weight, 37½ pounds.

LOAD.	INCH.		LOAD.	INCH.		LOAD.	INCH.	
Pounds.	Compression.	Set.	Pounds.	Compression.	Set.	Pounds.	Compression.	Set.
5,000	....	....	90,000	.0130	....	5,000	....	.0070
10,000	.0020	....	100,000	.0141	....	200,000	.0240	....
20,000	.0040	....	5,000	....	.0050	220,000	.0255	....
30,000	.0060	....	100,000	.0142	....	238,000	first crack.	
40,000	.0075	....	120,000	.0160	....	240,000	.0280	....
50,000	.0090	....	140,000	.0180	....	260,000	.0300	....
60,000	.0102	....	160,000	.0195	....	280,000	.0330	....
70,000	.0110	....	180,000	.0210	....	286,800	.0350 { snappi'g	{ sound.
80,000	.0122	....	200,000	.0230	....	301,100	broken	....

SPECIAL TABLE II.—(*Continued.*)

8-INCH CUBE, MARKED *Dc* ; BEDS NOT PLASTERED.

Actual size: Bed = 8″.03 × 8″.07; Height = 8″.00; Weight, 37½ pounds.

Load.	Inch.		Load.	Inch.		Load.	Inch.	
Pounds.	Compression.	Set.	Pounds.	Compression.	Set.	Pounds.	Compression.	Set.
5,000	....	....	100,000	.0100	....	200.000	.0190	....
10,000	.0010	....	120,000	.0118	....	220,000	.0207	....
20,000	.0020	....	140,000	.0133	....	240,000	.0227	....
40,000	.0045	....	160,000	.0150	....	260,000	.0250	....
60,000	.0065	....	180,000	.0166	....	280,000	.0282	....
80,000	.0082	....	200,000	.0182	....	285,000	.0296	....
100,000	.0100	....	5,000	....	.0025	294,100	broken	....
5,000	....	.0010	180,000	a corner off		....	....	....

8-INCH CUBE, MARKED *Dd* ; BEDS NOT PLASTERED.

Actual size: Bed = 8″.04 × 8″.00; Height = 8″.04; Weight, 39 pounds.

Load.	Inch.		Load.	Inch.		Load.	Inch.	
Pounds.	Compression.	Set.	Pounds.	Compression.	Set.	Pounds.	Compression.	Set.
5,000	....	....	180,000	.0145	....	310,000	.0260	....
10,000	.0005	....	200,000	.0160	....	315,000	.0264	....
20,000	.0020	....	5,000	....	.0020	320,000	.0270	....
40,000	.0040	....	200,000	.0160	....	325,000	.0280	....
60,000	.0062	....	220,000	.0175	....	330,000	.0288	....
80,000	.0077	....	240,000	.0190	....	335,000	.0292	....
100,000	.0090	....	260,000	.0203	....	340,000	.0300	....
5,000	....	.0010	280,000	.0223	....	345,000	.0305	....
100,000	.0092	....	285,000	.0230	....	350,000	.0310	begins to scale off.
120,000	.0105	....	290,000	.0239	....	355,000	.0323	....
140,000	.0120	....	295,000	.0244	....	358,000	.0335	....
160,000	.0130	....	300,000	.0250	....	360,000	broken	....

8-INCH CUBE, MARKED *De* ; BEDS NOT PLASTERED.

Actual size: Bed = 7″.98 × 8″.03; Height = 8″.02; Weight, 38 pounds.

Load.	Inch.		Load.	Inch.		Load.	Inch.	
Pounds.	Compression.	Set.	Pounds.	Compression.	Set.	Pounds.	Compression.	Set.
5,000	....	....	100,000	.0100	....	220,000	.0193	....
10,000	.0010	....	120,000	.0114	....	240,000	.0213	....
20,000	.0025	....	140,000	.0130	....	260,000	.0239	....
40,000	.0048	....	160,000	.0144	....	280,000	.0260	....
60,000	.0065	....	180,000	.0160	....	296,000	corner off	
80,000	.0080	....	200,000	.0178	....	299,200	broken	....
100,000	.0099	....	5,000	....	.0027	....	....	....
5,000	....	.0015	200,000	.0180	....	....	....	....

SPECIAL TABLE II.—(Continued.)

8-INCH CUBE, MARKED Df; BEDS NOT PLASTERED.
Actual size: Bed = 8".00 × 8".04; Height = 8".00; Weight, 39 pounds.

LOAD. Pounds.	INCH. Compression.	Set.	LOAD. Pounds.	INCH. Compression.	Set.	LOAD. Pounds.	INCH. Compression.	Set.
5,000	....	....	100,000	0100	......	220,000	0189	....
10,000	.0010	... .	120,000	.0112	....	240,000	.0205	..... .
20,000	.0028	....	140,000	.0127	....	260,000	.0224	....
40,000	.0050	....	160,000	.0140	....	280,000	.0242	....
60,000	.0067	....	180,000	.0133	. ...	300,000	.0270	....
80,000	.0082	....	200,000	.0171	....	304,000	cracked	....
100,000	.0098	....	5,000	....	.0021	310,000	.0290	....
5,000	....	.0012	200,000	.0172	....	338,000	broken	....

9-INCH CEMENT CUBE, MARKED Da; BEDS NOT PLASTERED.
Actual size: Bed = 9".05 × 9".01; Height = 9".04; Weight, 56 pounds.

LOAD. Pounds.	INCH. Compression.	Set.	LOAD. Pounds.	INCH. Compression.	Set.	LOAD. Pounds.	INCH. Compression.	Set.
5,000	....	....	140,000	.0154	....	300,000	.0260	....
10,000	.0012	....	160,000	.0168	....	5,000	....	.0050
20,000	.0032	....	180,000	.0180	....	300,000	.0265	....
40,000	.0063	....	200,000	.0191	....	320,000	.0280	....
60,000	.0090	....	5,000	....	.0032	340,000	.0292	....
80,000	.0110	....	200,000	.0192	....	345,000	begins to	crack
100,000	.0122	....	220,000	.0205	....	360,000	.0325	....
5,000	....	.0022	240,000	.0219	....	373,000	broken	....
100,000	.0125	....	260,000	.0230	....	....	....	....
120,000	.0140	....	280,000	.0244	....	....	....	....

9-INCH CEMENT CUBE, MARKED Db; BEDS NOT PLASTERED.
Actual size: Bed = 9".02 × 9".12; Height = 9".05; Weight, 56 pounds.

LOAD. Pounds.	INCH. Compression.	Set.	LOAD. Pounds.	INCH. Compression.	Set.	LOAD. Pounds.	INCH. Compression.	Set.
5,000	....	....	140,000	.0181	....	300,000	.0282	....
10,000	.0017	....	180,000	.0210	....	320,000	.0295	....
20,000	.0038	....	200,000	.0222	... .	327,000	corner	off
40,000	.0072	....	5,000	....	.0030	330,000	.0302	....
60,000	.0102	....	200,000	.0224	....	340,000	.0309	....
80,000	.0125	....	240,000	.0245	....	350,000	.0315	....
100,000	.0145	....	280,000	.0270	....	360,000	.0320	....
5,000	....	.0020	300,000	.0280	....	373,000	broken	....
100,000	.0150	....	5,000	....	.0045	....	....	....

SPECIAL TABLE II.—(Continued.)

9-Inch Cement Cube, marked Dc; Beds not Plastered.

Actual size: Bed = 9″.00 × 9″.00; Height = 8″.99; Weight, 55 pounds.

Load. Pounds.	Inch. Compression.	Set.	Load. Pounds.	Inch. Compression.	Set.	Load. Pounds.	Inch. Compression.	Set.
5,000	....	....	100,000	.0120	....	300,000	.0259	....
10,000	.0012	....	140,000	.0155	....	5,000	....	.0050
20,000	.0035	....	180,000	.0177	....	300,000	.0262	....
40,000	.0063	....	200,000	.0190	....	330,000	.0283	....
60,000	.0082	....	5,000	....	.0030	350,000	.0330	....
80,000	.0100	....	200,000	.0191	....	395,400	yielding s	uddenly
100,000	.0117	....	240,000	.0215	....	396,000	broken	....
5,000	....	.0015	280,000	.0243	....	....	....	....

9-Inch Cement Cube, marked Dd; Beds not Plastered.

Actual size: Bed = 9″.02 × 9″.04; Height = 9″.05; Weight, 56½ pounds.

Load. Pounds.	Inch. Compression.	Set.	Load. Pounds.	Inch. Compression.	Set.	Load. Pounds.	Inch. Compression.	Set.
5,000	....	....	100,000	.0071	....	300,000	.0202	....
10,000	.0005	....	140,000	.0094	....	5,000	....	.0030
20,000	.0011	....	180,000	.0120	....	300,000	.0210	....
40,000	.0030	....	200,000	.0132	....	340,000	.0240	⎰ slight
60,000	.0041	....	5,000	....	.0015	370,000	.0278	⎱ cracks
80,000	.0055	....	200,000	.0135	....	380,000	.0288	....
100,000	.0070	....	240,000	.0160	....	390,000	.0295	....
5,000	....	.0010	280,000	.0182	....	....	burst suddenly	

9-Inch Cement Cube, marked De; Beds not Plastered.

Actual size: Bed = 9″.07 × 9″.00; Height = 9″.03; Weight, 56 pounds.

Load. Pounds.	Inch. Compression.	Set.	Load. Pounds.	Inch. Compression.	Set.	Load. Pounds.	Inch. Compression.	Set.
5,000	....	....	140,000	.0120	....	300,000	.0215	....
10,000	.0020	....	180,000	.0142	....	340,000	.0240	....
20,000	.0030	....	200,000	.0155	....	370,000	.0260	....
40,000	.0052	....	5,000	....	.0030	380,000	.0270	....
60,000	.0070	....	200,000	.0155	....	390,000	.0275	....
80,000	.0081	....	240,000	.0178	....	400,000	.0284	....
100,000	.0095	....	280,000	.0200	....	458,600	begins to crack	
5,000	....	.0020	300,000	.0212	....	468,200	broken	....
100,000	.0096	....	5,000	....	.0040	....	....	....

SPECIAL TABLE II.—(*Continued.*)

9-Inch Cement Cube, marked *Df*; Beds not Plastered,
Actual size: Bed = 9".05 × 9".10; Height = 8".98; Weight, 55½ pounds.

Load.	Inch.		Load.	Inch.		Load.	Inch.	
Pounds.	Compression.	Set.	Pounds.	Compression.	Set.	Pounds.	Compression.	Set.
5,000	....	....	100,000	.0110	....	280,000	.0258	.....
10,000	.0010	....	130,000	cracked	....	300,000	.0288	....
20,000	.0030	....	140,000	.0142	....	5,000	....	.0081
40,000	.0052	....	180,000	.0172	....	300,000	.0308	....
60,000	.0071	....	200,000	.0190	....	310,000	.0328	....
80,000	.0090	....	5,000	.. ....	.0045	325,000	broken	....
100,000	.0108	....	200,000	.0192	....	....	....	....
5,000	....	.0020	240,000	.0220	....	....	....	....

10-Inch Cement Cube, marked *Da*; Beds not Plastered.
Actual size: Bed = 10".08 × 10".05; Height = 9".97; Weight, 75½ pounds.

Load.	Inch.		Load.	Inch.		Load.	Inch.	
Pounds.	Compression.	Set.	Pounds.	Compression.	Set.	Pounds.	Compression.	Set.
5,000	....	....	100,000	.0120	....	300,000	.0235	....
10,000	.0018	....	140,000	.0145	....	5,000	....	.0060
20,000	.0040	....	180,000	.0170	....	300,000	.0238	....
40,000	.0070	....	200,000	.0180	....	318,000	cracked	....
60,000	.0090	....	5,000	....	.0040	320,000	.0253	....
80,000	.0110	....	200,000	.0180	....	340,000	.0267	cracking
100,000	.0120	....	240,000	.0200	....	351,000	.0292	....
5,000	....	.0022	280,000	.0221	....	395,300	broken	....

10-Inch Cement Cube, marked *Db*; Beds not Plastered.
Actual size: Bed = 10".02 × 10".00; Height × 10".00; Weight, 76½ pounds.

Load.	Inch.		Load.	Inch.		Load.	Inch.	
Pounds.	Compression.	Set.	Pounds.	Compression.	Set.	Pounds.	Compression.	Set.
5,000	....	....	180,000	.0088	... .	380,000	.0180	... ...
10,000	.0003	....	200,000	.0098	... .	400,000	.0192	....
20,000	.0010	... ...	5,000	....	.0010	5,000	....	.0020
40,000	.0022	....	200,000	.0099	....	400,000	.0192	....
60,000	.0031	....	240,000	.0115	....	440,000	.0218	....
80,000	.0040	....	280,000	.0132	....	460,000	.0230	....
100,000	.0050	....	300,000	.0145	....	470,000	.0240	....
5,000	....	.0010	5,000	... .	.0015	480,000	.0250	....
100,000	.0052	....	300,000	.0145	....	540,000	cracked	....
140,000	.0072	....	340,000	.0160	....	587,100	broken	....

SPECIAL TABLE II.—(*Continued.*)

10-Inch Cement Cube, marked *Dc*; Beds not Plastered.
Actual size: Bed = 10".09 × 10".04; Height = 10".00; Weight, 76½ pounds.

Load. Pounds.	Inch. Compression.	Set.	Load. Pounds.	Inch. Compression.	Set.	Load. Pounds.	Inch. Compression.	Set.
5,000	....	....	140,000	.0148	....	300,000	.0220	....
10,000	.0008	....	180,000	.0165	....	340,000	.0240	....
20,000	.0042	....	200,000	.0174	....	380,000	.0260	....
40,000	.0075	....	5,000	....	.0035	400,000	.0270	....
60,000	.0100	.. ...	200,000	.0175	....	5,000	....	.0071
80,000	.0115	....	240,000	.0190	....	400,000	.0275	....
100,000	.0128	....	280,000	.0210	... .	440,000	.0288	....
5,000	....	.0025	300,000	.0220	....	460,000	.0301	....
100,000	.0128	....	5,000	....	.0048	519,000	broken	....

10-Inch Cement Cube, marked *Dd*; Beds not Plastered.
Actual size: Bed = 10".08 × 10".10; Height = 10".00; Weight, 77 pounds.

Load. Pounds.	Inch. Compression.	Set.	Load. Pounds.	Inch. Compression.	Set.	Load. Pounds.	Inch. Compression.	Set.
5,000	....	....	5,000	....	.0010	240,000	.0202	....
10,000	.0012	....	100,000	.0120	....	280,000	.0225	....
20,000	.0030	....	140,000	.0148	....	300,000	.0238	....
40,000	.0062	....	180,000	.0158	....	5,000	....	.0041
60,000	.0084	....	200,000	.0180	....	300,000	.0252	....
80,000	.0102	....	5,000	....	.0020	320,000	.0273	....
100,000	.0120	....	200,000	.0182	....	430,100	broken	....

10 Inch Cement Cube, marked *De*; Beds not Plastered.
Actual size: Bed = 10".00 × 10".05; Height = 10".08; Weight, 76 pounds.

Load. Pounds.	Inch. Compression.	Set.	Load. Pounds.	Inch. Compression.	Set.	Load. Pounds.	Inch. Compression.	Set.
5,000	....	....	140,000	.0116	....	300,000	.0220	....
10,000	.0008	....	180,000	.0140	....	340,000	.0242	....
20,000	.0020	....	200,000	.0150	....	380,000	.0272	....
40,000	.0042	....	5,000	....	.0020	400,000	.0290	....
60,000	.0062	....	240,000	.0172	....	5,000	....	.0060
80,000	.0075	....	242,000	side cracked		400,000	.0300	....
100,000	.0090	....	280,000	.0202	....	473,400	broken	....
5,000	....	.0010	300,000	.0218	....			
100,000	.0090	....	5,000	....	.0032			

SPECIAL TABLE II.—(*Continued.*)

10-INCH CEMENT CUBE, MARKED *Df*; BEDS NOT PLASTERED.

Actual size: Bed = 10".01 × 10".05; Height = 9".99; Weight, 76½ pounds.

LOAD.	INCH.		LOAD.	INCH.		LOAD.	INCH.	
Pounds.	Compression.	Set.	Pounds.	Compression.	Set.	Pounds.	Compression.	Set.
5,000	....	....	180,000	.0130	....	380,000	.0245	....
10,000	.0010	....	200,000	.0140	....	400,000	.0260	....
20,000	.0020	....	5,000	....	.0015	5,000	....	.0042
40,000	.0040	....	200,000	.0140	....	400,000	.0265	....
60,000	.0053	....	240,000	.0160	....	420,000	.0280	....
80,000	.0065	....	280,000	.0180	....	440,000	.0290	....
100,000	.0078	....	300,000	.0193	....	460,000	.0304	....
5,000	....	.0010	5,000	....	.0025	472,000	.0320	....
100,000	.0078	....	300,000	.0198	....	477,600	broken	....
140,000	.0102	....	340,000	.0220	....	....	....	....

11-INCH CEMENT CUBE, MARKED *Da*; BEDS NOT PLASTERED.

Actual size: Bed = 11".00 × 11".15; Height = 11".00; Weight, 101 pounds.

LOAD.	INCH.		LOAD.	INCH.		LOAD.	INCH.	
Pounds.	Compression.	Set.	Pounds.	Compression.	Set.	Pounds.	Compression.	Set.
5,000	....	....	....	....	....	400,000	.0240	....
10,000	.0005	....	5,000	....	.0012	440,000	.0260	....
20,000	.0020	....	200,000	.0122	....	480,000	.0290	....
40,000	.0035	....	240,000	.0141	....	500,000	.0302	....
60,000	.0048	....	280,000	.0160	....	5,000	....	.0052
80,000	.0060	....	300,000	.0175	....	500,000	.0313	....
100,000	.0070	....	5,000	....	.0020	510,000	.0324	....
5,000	....	.0008	300,000	.0175	....	520,000	.0332	....
100,000	.0070	....	340,000	.0200	....	530,000	.0340	cracks
140,000	.0092	....	380,000	.0220	....	540,000	.0350	....
180,000	.0110	....	400,000	.0235	....	591,200	broken	....
200,000	.0120	....	5,000	....	.0032			

11

SPECIAL TABLE II.—(*Continued.*)

11-Inch Cement Cube, marked *Db*; Beds Plastered.

Actual size: Bed = 11".05 × 11".00; Height = 11".00 (or 11".03 including plaster); Weight, 100 pounds.

Load.	Inch.		Load.	Inch.		Load.	Inch.	
Pounds.	Compression.	Set.	Pounds.	Compression.	Set.	Pounds.	Compression.	Set.
5,000	....	....	280,000	.0155	....	510,000	.0302	....
10,000	.0008	....	300,000	.0165	....	520,000	.0312	....
20,000	.0022	....	5,000	....	.0040	530,000	.0320	....
40,000	.0042	....	300,000	.0168	....	540,000	.0325	....
80,000	.0065	....	340,000	.0188	....	550,000	.0330	....
100,000	.0073	....	380,000	.0210	....	560,000	.0338	....
5,000	....	.0020	400,000	.0220	....	570,000	.0350	corner cracked
100,000	.0073	....	5,000	....	.0058	580,000	.0358	....
140,000	.0090	....	400,000	.0222	....	590,000	.0375	....
180,000	.0110	....	440,000	.0248	....	600,000	.0379	....
200,000	.0120	....	480,000	.0270	....	610,000	.0390	....
5,000	....	.0031	500,000	.0288	....	620,000	.0402	....
200,000	.0120	....	5,000	....	.0078	630,000	.0415	....
240,000	.0138	....	500,000	.0292	....	633,000	.0430	broken

11-Inch Cement Cube, marked *Dc*; Beds Plastered.

Actual size: Bed = 11".00 × 11".18; Height = 11".00 (or 11".02 including plaster); Weight, 101½ pounds.

Load.	Inch.		Load.	Inch.		Load.	Inch.	
Pounds.	Compression.	Set.	Pounds.	Compression.	Set.	Pounds.	Compression.	Set.
5,000	....	....	300,000	.0162	....	600,000	.0372	....
10,000	.0005	....	400,000	.0220	....	610,000	.0380	....
20,000	.0015	....	5,000	....	.0037	620,000	.0387	....
40,000	.0030	....	400,000	.0225	....	630,000	.0395	....
80,000	.0050	....	500,000	.0285	....	640,000	.0405	....
100,000	.0062	....	5,000	....	.0050	650,000	.0422	....
5,000	....	.0010	500,000	.0290	....	660,000	.0428	....
100,000	.0062	....	520,000	.0302	....	670,000	.0440	....
200,000	.0112	....	540,000	.0320	....	680,000	.0460	....
5,000	....	.0019	560,000	.0332	....	690,000	.0470	....
200,000	.0110	....	570,000	.0342	....	700,000	.0480	....
300,000	.0160	....	580,000	.0352	....	725,100	broken	....
5,000	....	.0025	590,000	.0362	....	....	....	....

SPECIAL TABLE II.—(*Continued.*)

11-INCH CEMENT CUBE, MARKED *Dd* ; BEDS PLASTERED.

Actual size: Bed = 11″.03 × 11″.21; Height = 11″.00 (or 11″.02 including plaster); Weight, 101½ pounds.

LOAD.	INCH.		LOAD.	INCH.		LOAD.	INCH.	
Pounds.	Compression.	Set.	Pounds.	Compression.	Set.	Pounds.	Compression.	Set.
5,000	....	......	300,000	.0135	....	540,000	.0282	......
10,000	.0002	....	5,000	....	.0020	550,000	.0290	......
20,000	.0010	....	300,000	.0138	......	560,000	.0292	......
40,000	.0020	....	400,000	.0182	....	570,000	.0300	......
80,000	.0040	....	5,000	....	.0030	580,000	.0308	......
100,000	.0048	....	400,000	.0186	....	590,000	.0315	......
5,000	....	.0010	500,000	.0240	....	600,000	.0320	......
100,000	.0050	....	5,000	....	.0040	620,000	.0340	......
200,000	.0090	....	500,000	.0250	....	640,000	.0365	......
5,000	....	.0018	520,000	.0265	......	660,000	.0390	......
200,000	.0090	....	530,000	.0272	....	674,000	broken	......

11-INCH CEMENT CUBE, MARKED *De* ; BEDS PLASTERED.

Actual size: Bed = 11″.02 × 11″.21; Height = 10″.99 (or 11″.02 including plaster); Weight, 101 pounds.

LOAD.	INCH.		LOAD.	INCH.		LOAD.	INCH.	
Pounds.	Compression.	Set.	Pounds.	Compression.	Set.	Pounds.	Compression.	Set.
5,000	....	....	300,000	.0140	....	540,000	.0292	......
10,000	.0008	....	5,000	....	.0025	560,000	.0310	......
20,000	.0013	......	300,000	.0140	....	580,000	.0325	......
40,000	.0028	....	400,000	.0190	......	600,000	.0340	......
80,000	.0045	....	5,000	....	.0032	620,000	.0360	......
100,000	.0052	....	400,000	.0190	......	640,000	.0382	......
5,000	....	.0010	500,000	.0250	....	660,000	.0408	cracks in sight
100,000	.0060	....	5,000	....	.0049	680,000	.0432	
200,000	.0095	....	500,000	.0268	....	690,200	broken	......
5,000	....	.0018	510,000	.0273	......			
200,000	.0097	....	520,000	.0282	....			

SPECIAL TABLE II.—(*Continued.*)

11-INCH CEMENT CUBE, MARKED *Df*; BEDS PLASTERED.

Actual size: Bed = 11".05 × 11".05; Height = 11".02 (or 11".04 including plaster); Weight, 100 pounds.

Load.	Inch.		Load.	Inch.		Load.	Inch.	
Pounds.	Compression.	Set.	Pounds.	Compression.	Set.	Pounds.	Compression.	Set.
5,000	....	....	200,000	.0112	....	520,000	.0290	......
10,000	.0008	....	300,000	.0160	....	540,000	.0300	......
20,000	.0019	....	5,000	....	.0038	560,000	.0315	......
40,000	.0035	....	300,000	.0160	....	580,000	.0330	......
80,000	.0058	....	400,000	.0210	....	600,000	.0350	......
100,000	.0069	... .	5,000	....	.0050	620,000	.0372	......
5,000	....,...	.0020	400,000	.0212	....	630,000	.0390	cracks
100,000	.0069	....	500,000	.0270	....	640,000	.0410	......
200,000	.0110	....	5,000	....	.0069	645,600	.0422	broken
5,000	....	.0028	500,000	.0280	....	....	....	......

12-INCH CEMENT CUBE, MARKED *Da*; BEDS PLASTERED.

Actual size: Bed = 12".05 × 12".00; Height = 12".00, exclusive of plaster; Weight, 129 pounds.

Load.	Inch.		Load.	Inch.		Load.	Inch.	
Pounds.	Compression.	Set.	Pounds.	Compression.	Set.	Pounds.	Compression.	Set.
5,000	....	....	5,000	....	.0028	600,000	.0330	.. ....
40,000	.0022	....	300,000	.0135	....	620,000	.0352	......
80,000	.0040	....	400,000	.0182	......	640,000	.0370	......
100,000	.0048	....	5,000	....	.0038	660,000	.0390	cracks in sight
5,000	... .	.0010	400,000	.0182	....	680,000	.0422	......
100,000	.0048	....	500,000	.0240	....	690,000	.0450	......
200,000	.0088	....	5,000	....	.0050	700,000	.0475	......
5,000	....	.0020	500,000	.0248	....	710,000	.0520	broken
200,000	.0090	....	600,000	.0320	....	....	....	......
300,000	.0132	....	5,000	....	.0080	....	....	......

SPECIAL TABLE II.—(*Continued.*)

12-INCH CEMENT CUBE, MARKED Db; BEDS PLASTERED.

Actual size: Bed = 12".08 × 12".05; Height = 11".97, exclusive of plaster; Weight, 129 pounds.

LOAD.	INCH.		LOAD.	INCH.		LOAD.	INCH.	
Pounds.	Compression.	Set.	Pounds.	Compression.	Set.	Pounds.	Compression.	Set.
5,000	....	....	300,000	.0159	....	5,000	....	.0071
40,000	.0039	....	5,000	....	.0030	600,000	.0330	....
80,000	.0060	....	400,000	.0200	....	640,000	.0366	....
100,000	.0070	....	5,000	....	.0039	673,000	.0402 }	cracks in sight
5,000	....	.0017	500,000	.0260	....	783,000	broken	....
200,000	.0115	....	5,000	....	.0050	....	....	....
5,000	....	.0022	600,000	.0320	....	....	....	....

12-INCH CEMENT CUBE, MARKED Dc; BEDS PLASTERED.

Actual size: Bed = 12".00 × 12".03; Height = 12".03, exclusive of plaster; Weight, 130½ pounds.

LOAD.	INCH.		LOAD.	INCH.		LOAD.	INCH.	
Pounds.	Compression.	Set.	Pounds.	Compression.	Set.	Pounds.	Compression.	Set.
5,000	....	....	5,000	....	.0050	750,000	.0390	....
40,000	.0030	....	600,000	.0270	....	760,000	.0400	....
80,000	.0049	....	5,000	....	.0060	800,000	....	....
100,000	.0058	....	600,000	.0280	....	5,000	{ Remaining eight	
5,000	....	.0022	640,000	.0300	....	800,000	{ minutes.	
200,000	.0100	....	660,000	.0315	....	5,000	{ Remaining eight	
5,000	....	.0030	680,000	.0330	....	800,000	{ minutes.	
300,000	.0140	....	700,000	.0345 }	cracks in sight	5,000	{ Remaining eight	
5,000	....	.0035	710,000	.0352	....	800,000	{ minutes.	
400,000	.0180	....	720,000	.0365	....	5,000	{ Remaining eight	
5,000	....	.0040	730,000	.0372	....	800,000	{ minutes.	
500,000	.0220	....	740,000	.0382	....	5,000	{ Failed rapidly	
						800,000	{ and broke	

SPECIAL TABLE II.—(*Continued.*)

12-INCH CEMENT CUBE, MARKED *Dd*; BEDS PLASTERED.

Actual size: Bed = 12".10 × 11".30; Height = 12".00 (or 12".03 including plaster); Weight, 123 pounds.

Load. Pounds.	Inch. Compression.	Set.	Load. Pounds.	Inch. Compression.	Set.	Load. Pounds.	Inch. Compression.	Set.
5,000	.... ..	....	5,000	.. ..	.0090	770,000	.0458	....
40,000	.0025	....	600,000	.0306	....	780,000	.0468	... ...
80,000	.0042	....	620,000	.0322	....	790,000	.0475	....
100,000	.0052	....	640,000	.0338	....	798,000	.0488	{ small pieces fly off
5,000	....	.0030	660,000	.0352	....	800,000		)
200,000	.0100		680,000	.0370	....	5,000		Each application
5,000	....	.0040	700,000	.0385	....	800,000		of the maxi-
300,000	.0140	. ...	708,000	cracks in	sight	5,000		mum load
5,000	....	.0045	710,000	.0392	....	800,000		caused small pieces to fly off,
400,000	.0180	....	720,000	.0405	....	5,000		and increased
5,000	....	.0052	730,000	.0414	....	800,000		the size of cracks.
500,000	.0232	....	740,000	.0422	....	5,000		)
5,000	....	.0065	750,000	.0437	....	800,000		When this load had been maintained about 6 minutes, the piece rapidly yielded and broke.
600,000	.0300	....	760,000	.0450	....			

12-INCH CEMENT CUBE, MARKED *De*; BEDS PLASTERED.

Actual size: Bed = 12".05 × 12".00; Height = 12".00 (or 12".07 including plaster); Weight, 131 pounds.

Load. Pounds.	Inch. Compression.	Set.	Load. Pounds.	Inch. Compression.	Set.	Load. Pounds.	Inch. Compression.	Set.
5,000	.. ..	....	600,000	.0285	....	200,000	.0215	....
40,000	.0025	....	620,000	.0302	....	5,000	....	.0132
80,000	.0042	....	640,000	.0318	....	300,000	.0250	....
100,000	.0050	....	660,000	.0330	....	5,000	....	.0132
5,000	....	.0000	680,000	.0345	....	400,000	.0290	....
200,000	.0085	....	700,000	.0357	....	5,000	....	.0132
5,000	....	.0010	720,000	.0370	....	500,000	.0325	....
300,000	.0125	....	740,000	.0382	....	5,000	....	.0133
5,000	....	.0020	760,000	.0400 { pieces fly off	....	600,000	.0365	....
400,000	.0170	....	770,000		....	5,000	....	.0135
5,000	....	.0032	780,000	.0420	....	700,000	.0410	... ...
500,000	.0220	....	800,000	.0445	....	5,000	....	.0140
5,000	....	.0050	5,000	....	.0137	770,000	{ pieces fly off	....
600,000	.0275	....	100,000	.0175	....:..	800,000	Sustained this load about ½ minute, then failed rapid- ly, and broke.
5,000	....	.0070	5,000	....	.0132			

SPECIAL TABLE II.—(*Continued.*)

12-Inch Cement Cube, marked *Df*; Beds Plastered.

Actual size: Bed = 12".00 × 12".06; Height = 12".00 (or 12".01 including plaster); Weight, 130 pounds.

Load.	Inch.		Load.	Inch.		Load.	Inch.	
Pounds.	Compression.	Set.	Pounds.	Compression.	Set.	Pounds.	Compression.	Set.
5,000	....	....	5,000	....	.0040	640,000	.0340	....
40,000	.0030	....	400,000	.0185	....	660,000	.0355	....
80,000	.0050	....	5,000	....	.0050	680,000	.0372	....
100,000	.0060	....	500,000	.0240	....	685,000	{ pieces fly off.	....
5,000	....	.0020	5,000	....	.0065	700,000	.0410 { decided yielding,	....
200,000	.0100	....	600,000	.0302	....	715,500	{ fragments flying off.	
5,000	....	.0030	5,000	....	.0085	773,200	broken	
300,000	.0142	....	620,000	.0328	....	....		

PIERS OF PRISMS OF NEAT (DYCKERHOFF) CEMENT.

THREE PRISMS, EACH 12 INCHES SQUARE, 6 INCHES HIGH; BEDS PLASTERED; DRY JOINTS.

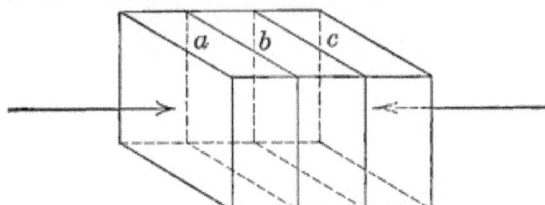

18".06, including plaster.

Actual size: Prism *a*—Bed = 12".01 × 12".04; Height = 5".98; Weight, 64 pounds, 12 ounces.
Prism *b*—Bed = 12".05 × 11".99; Height = 5".94; Weight, 64 pounds, 8 ounces.
Prism *c*—Bed = 12".13 × 12".08; Height = 5".95; Weight, 64 pounds, 14 ounces.

Load.	Inch.		Load.	Inch.		Load.	Inch.	
Pounds.	Compression.	Set.	Pounds.	Compression.	Set.	Pounds.	Compression.	Set.
5,000	....	....	240,000	.0201	....	500,000	.0498	....
10,000	.0015	....	260,000	.0220	....	5,000	....	.0222
20,000	.0030	....	280,000	.0232	....	500,000	.0512	....
40,000	.0055	....	300,000	.0252	....	520,000	.0540	....
60,000	.0072	....	5,000	....	.0091	540,000	.0565	....
80,000	.0088	....	300,000	.0252	....	560,000	.0590	....
100,000	.0101	....	320,000	.0270	....	580,000	.0628	....
5,000	....	.0039	340,000	.0290	....	600,000	.0660	....
100,000	.0102	....	360,000	.0310	....	5,000	....	.0322
120,000	.0120	....	380,000	.0335	....	600,000	.0678	....
140,000	.0130	....	400,000	.0360	....	620,000	.0700	....
160,000	.0148	....	5,000	....	.0147	640,000	.0730	....
180,000	.0160	....	400,000	.0360	....	660,000	.0765	{ snappi'g sound.
200,000	.0175	....	420,000	.0389	....	680,000	.0800	....
5,000	....	.0062	440,000	.0412	....	700,000	.0840	{ cracks at joint *a—b.*
200,000	.0178	....	460,000	.0445	....	5,000	....	.0420
220,000	.0190	....	480,000	.0475	....	690,000	{ failed rapidly and broke.	

SPECIAL TABLE II.—(*Concluded.*)

THREE PRISMS, EACH 12 INCHES SQUARE, 8 INCHES HIGH; BEDS PLASTERED; DRY JOINTS.

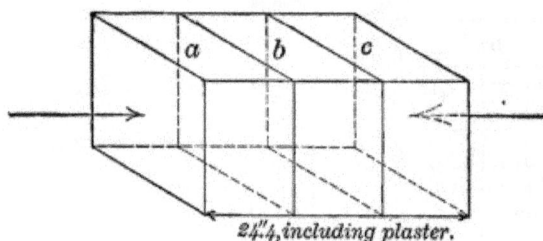

Actual size: Prism a—Bed $= 12''.03 \times 12''.14$; Height $= 8''.09$; Weight, 86 pounds, — ounces.
             Prism b—Bed $= 11''.98 \times 12''.08$; Height $= 8''.08$; Weight, 86 pounds, 12 ounces.
             Prism c—Bed $= 12''.08 \times 12''.10$; Height $= 8''.08$; Weight, 86 pounds, 8 ounces.

| LOAD. | INCH. | | LOAD. | INCH. | | LOAD. | INCH. | |
Pounds.	Compression.	Set.	Pounds.	Compression.	Set.	Pounds.	Compression.	Set.
5,000	......	......	220,000	.0232	......	460,000	.0471	......
10,000	.0012	......	240,000	.0250	......	480,000	.0500	......
20,000	.0032	......	260,000	.0267	......	500,000	.0520	......
40,000	.0071	......	280,000	.0282	......	5,000	......	.0162
60,000	.0096	......	300,000	.0300	......	500,000	.0530	......
80,000	.0114	......	5,000	......	.0088	520,000	.0560	......
100,000	.0130	......	300,000	.0302	......	540,000	.0590	......
5,000	......	.0042	320,000	.0320	......	560,000	.0615	......
100,000	.0132	......	340,000	.0340	......	580,000	.0645	began to flake at joint a–b
120,000	.0149	......	360,000	.0360	......	600,000	.0688	... ...
140,000	.0162	......	380,000	.0380	......	5,000	......	.0232
160,000	.0180	......	400,000	.0400	......	600,000	.0702	......
180,000	.0200	......	5,000	......	.0120	620,000	.0735	......
200,000	.0215	... .	400,000	.0402	......	640,000	.0765	......
5,000	......	.0062	420,000	.0425	......	654,800	.0820	failed
200,000	.0217	......	440,000	.0448	......	suddenly under this load.		

A continuous longitudinal seam opened along the three prisms, splitting off one corner of the pier; other similar seams also opened. The main fragment of prism a was of pyramidal form, with steep side slopes; prisms b and c were broken up in longitudinal fragments, about parallel to the line of pressure.

SPECIAL TABLE III.

SHOWING AMOUNT OF COMPRESSION AND SET OF CUBES OF CONCRETE.

Composition: 1 vol. Newark Company's Rosendale Cement, 3 vols. Sand, 2 vols. Gravel, 4 vols. Broken Stone.

10-INCH CONCRETE CUBE, MARKED *F6*; BEDS PLASTERED.

Actual size: Bed = 10″.11 × 10″.04; Height = 10″.16 (or 10″.22 including plaster): Weight, 78 pounds.

LOAD.	INCH.		LOAD.	INCH.		LOAD.	INCH.	
Pounds.	Compression.	Set.	Pounds.	Compression.	Set.	Pounds.	Compression.	Set.
5,000	....	... ...	50,000	.0082	....	85,000	.0200	....
10,000	.0010	....	5,000	....	.0051	90,000	.0230	... ...
15,000	.0020	....	50,000	.0088	....	95,000	.0270	....
20,000	.0030	....	35,000	.0097	....	100,000	.0320	....
25,000	.0040	... ...	60,000	.0110	....	105,000	.0385	....
30,000	.0048	....	65,000	.0120	....	110,000	.0500	....
35,000	.0058	....	70,000	.0140	....	115,000	.0670	....
40,000	.0065	....	75,000	.0155	....	120,000	.1000	broken
45,000	.0075	....	80,000	.0188	....	Surface cracks appeared immediately before the ultimate load was reached.		

12-INCH CONCRETE CUBE, MARKED *F6*; BEDS PLASTERED.

Actual size: Bed = 12″.06 × 12″.04; Height = 12″.00 (or 12″.02 including plaster): Weight, 136 pounds.

LOAD.	INCH.		LOAD.	INCH.		LOAD.	INCH.	
Pounds.	Compression.	Set.	Pounds.	Compression.	Set.	Pounds.	Compression.	Set.
5,000	. ...	... .	60,000	.0100	....	115,000	.0250	....
10,000	.0010	. ...	65,000	.0110	....	120,000	.0270	....
15,000	.0025	....	70,000	.0120	....	125,000	.0234	....
20,000	.0035	....	75,000	.0130	....	130,000	.0335	....
25,000	.0042	....	80,000	.0140	....	135,000	.0368	....
30,000	.0052	... .	85,000	.0150	....	140,000	.0420	....
35,000	.0060	....	90,000	.0160	....	145,000	.0480	... ...
40,000	.0070	....	95,000	.0175	....	150,000	.0560	cracks developing.
45,000	.0075	....	100,000	.0190	....	155,000	.0680	....
50,000	.0080	....	5,000	....	.0125	160,000	.0950	....
5,000	....	.0048	100,000	.0210	....	161,600	broken	....
50,000	.0082	....	105,000	.0225	....	....	....	....
55,000	.0092	....	110,000	.0240	....	....	....	....

SPECIAL TABLE III.—(*Continued.*)

14-INCH CONCRETE CUBE, MARKED *Fb* ; BEDS PLASTERED.

Actual size: Bed = 14".09 × 14".05; Height = 14".04 (or 14".13 including plaster); Weight, 211 pounds.

LOAD. Pounds.	INCH. Compression.	Set.	LOAD. Pounds.	INCH. Compression.	Set.	LOAD. Pounds.	INCH. Compression.	Set.
5,000	....	....	50,000	.0080	....	100,000	.0220	....
10,000	.0015	....	60,000	.0092	....	110,000	.0275	....
20,000	.0030	....	70,000	.0110	....	120,000	.0350	....
30,000	.0042	....	80,000	.0125	....	130,000	.0490	....
40,000	.0060	....	90,000	.0160	....	140,000	.0720	....
50,000	.0070	....	100,000	.0200	....	147,000	cracks in sight.	
5,000	....	.0045	5,000	....	.0130	148,000	broken	

16-INCH CONCRETE CUBE, MARKED *Fb* ; BEDS PLASTERED.

Actual size: Bed = 16".05 × 16".10; Height = 16".04 (or 16".16 including plaster); Weight, 325½ pounds.

LOAD. Pounds.	INCH. Compression.	Set.	LOAD. Pounds.	INCH. Compression.	Set.	LOAD. Pounds.	INCH. Compression.	Set.
5,000	....	....	100,000	.0083	....	180,000	.0235	....
10,000	.0012	....	5,000	....	.0042	190,000	.0275	... ...
20,000	.0028	....	100,000	.0090	....	200,000	.0325	....
30,000	.0035	....	110,000	.0100	... ...	210,000	.0385	....
40,000	.0040	.. ..	120,000	.0110	....	220,000	.0440	....
50,000	.0048	....	130,000	.0120	....	230,000	.0500	....
5,000	....	.0028	140,000	.0135	....	240,000	.0605	....
50,000	.0049	....	150,000	.0150	....	250,000	.0720	cracks in sight
60,000	.0052	....	5,000	....	.0080	260,000	.0920	
70,000	.0062	....	150,000	.0162	....	268,400	broken	....
80,000	.0069	.. ..	160,000	.0175	....	....	....	....
90,000	.0075	....	170,000	.0210	....	. ...	....	....

SPECIAL TABLE III.—(*Concluded.*)

18-INCH CONCRETE CUBE, MARKED *Fb*; BEDS PLASTERED.

Actual size: Bed = 18".00 × 17".62; Height = 18".00 (or 18".19 including plaster); Weight, 455 pounds.

LOAD.	INCH.		LOAD.	INCH.		LOAD.	INCH.	
Pounds.	Compression.	Set.	Pounds.	Compression.	Set.	Pounds.	Compression.	Set.
5,000	....	....	140,000	.0120	....	250,000	.0290	....
10,000	.0004	....	160,000	.0140	....	260,000	.0320	....
20,000	.0015	....	180,000	.0162	....	270,000	.0360	....
40,000	.0040	....	200,000	.0190	....	280,000	.0410	....
60,000	.0060	....	5,000	....	.0200	290,000	.0455	....
80,000	.0072	....	200,000	.0212	....	300,000	.0520	....
100,000	.0090	....	210,000	.0220	....	310,000	.0615	....
5,000	....	.0045	220,000	.0230	....	320,000	.0695	....
100,000	.0092	....	230,000	.0240	....	330,000	.0808	....
120,000	.0105	....	240,000	.0260	....	331,000	.0930	broken

---

SPECIAL TABLE IV.

SHOWING AMOUNT OF COMPRESSION AND SET OF CUBES OF MORTAR MADE WITH NORTON'S CEMENT.

Composition: 1 vol. Cement Paste, 1½ vols. Sand.

8-INCH MORTAR CUBE, MARKED *Aa*; BEDS PLASTERED.

Actual size: Bed = 8".05 × 8".03; Height = 8".11 (or 8".14 including plaster); Weight, 37 pounds.

LOAD.	INCH.		LOAD.	INCH.		LOAD.	INCH.	
Pounds.	Compression.	Set.	Pounds.	Compression.	Set.	Pounds.	Compression.	Set.
1,000	....	....	40,000	.0070	....	1,000	....	.0045
5,000	.0015	....	45,000	.0075	....	75,000	.0142	....
10,000	.0020	....	50,000	.0082	....	80,000	.0152	....
15,000	.0030	....	1,000	....	.0022	85,000	.0165	....
20,000	.0038	....	50,000	.0088	....	90,000	.0180	....
25,000	.0042	....	55,000	.0095	....	95,000	.0200	....
1,000	....	.0010	60,000	.0105	....	100,000	.0222	....
25,000	.0042	....	65,000	.0115	....	105,000	.0252	cracks
30,000	.0050	....	70,000	.0122	....	106,000	.0290	broken
35,000	.0060	....	75,000	.0138	....	....	....	....

SPECIAL TABLE IV.—(*Continued.*)

8-INCH MORTAR CUBE, MARKED *Ab*; BEDS PLASTERED.

Actual size: Bed = 8".05 × 8".05; Height = 7".99 (or 8".00 including plaster); Weight, 36¾ pounds.

LOAD. Pounds.	INCH. Compression.	Set.	LOAD. Pounds.	INCH. Compression.	Set.	LOAD. Pounds.	INCH. Compression.	Set.
1,000	....	....	40,000	.0085	....	1,000	....	.0052
5,000	.0020	....	45,000	.0095	....	75,000	.0152	....
10,000	.0030	....	50,000	.0102	....	80,000	.0160	....
15,000	.0042	....	1,000	....	.0035	85,000	.0172	....
20,000	.0050	....	50,000	.0105	....	95,000	.0200	....
25,000	.0062	....	55,000	.0110	....	100,000	.0210	....
1,000	....	.0010	60,000	.0120	....	105,000	.0230	....
25,000	.0062	....	65,000	.0130	....	110,000	.0252	....
30,000	.0070	....	70,000	.0140	....	115,000	.0280	....
35,000	.0076	....	75,000	.0150	....	120,000	.0353	broken

NOTE.—Cracks appeared when the load had reached 118,000 pounds.

12-INCH MORTAR CUBE, MARKED *Aa*; BEDS PLASTERED.

Actual size: Bed = 12".03 × 12".07; Height = 12".03 (or 12".24 including plaster); Weight, 118½ pounds.

LOAD. Pounds.	INCH. Compression.	Set.	LOAD. Pounds.	INCH. Compression.	Set.	LOAD. Pounds.	INCH. Compression.	Set.
5,000	....	....	80,000	.0753	....	150,000	.1032	....
10,000	.0010	....	90,000	.0800	....	160,000	.1075	....
20,000	.0045	....	100,000	.0845	....	170,000	.1125	....
30,000	.0222	....	5,000	....	.0760	180,000	.1185	....
40,000	.0420	....	100,000	.0870	....	190,000	.1260	....
50,000	.0555	....	110,000	.0890	....	192,000	.1330	broken
5,000	....	.0520	120,000	.0920	....	The plaster coating was rather soft and yielding, and comparatively thick, which may account for the observed rate of compression and set.		
50,000	.0556	....	130,000	.0960	....			
60,000	.0630	....	140,000	.0990	....			

SPECIAL TABLE IV.—(*Continued.*)

12-Inch Mortar Cube, marked *A b*; Beds Plastered.

Actual size: Bed = 12".02 x 12".02; Height = 12".17 (or 12".11 including plaster); Weight, 118¾ pounds.

Load.	Inch.		Load.	Inch.		Load.	Inch.	
Pounds.	Compression.	Set.	Pounds.	Compression.	Set.	Pounds.	Compression.	Set.
5,000	....	....	80,000	.0100	....	5,000	....	.0125
10,000	.0012	....	90,000	.0112	....	150,000	.0260	....
20,000	.0030	....	100,000	.0130	....	160,000	.0295	....
30,000	.0042	....	5,000	....	.0060	170,000	.0330	....
40,000	.0052	....	100,000	.0135	....	180,000	.0390	....
50,000	.0062	....	110,000	.0150	....	190,000	.0480	....
5,000	....	.0030	120,000	.0170	....	196,100	cracks in sight.	
50,000	.0065	....	130,000	.0192	....	197,400	broken	
60,000	.0075	....	140,000	.0220	....			
70,000	.0088	....	150,000	.0250	....			

16-Inch Mortar Cube, marked *A a*; Beds Plastered.

Actual size: Bed = 16".00 x 16".01; Height = 16".05 (or 16".13 including plaster); Weight, 284 pounds.

Load.	Inch.		Load.	Inch.		Load.	Inch.	
Pounds.	Compression.	Set.	Pounds.	Compression.	Set.	Pounds.	Compression.	Set.
5,000	....	....	5,000	....	.0030	240,000	.0240	....
10,000	.0010	....	100,000	.0100	....	250,000	.0258	....
20,000	.0022	....	120,000	.0110	....	260,000	.0280	....
30,000	.0038	....	140,000	.0128	....	270,000	.0292	....
40,000	.0048	....	160,000	.0145	....	280,000	.0310	....
50,000	.0056	....	180,000	.0160	....	290,000	.0340	....
60,000	.0068	....	200,000	.0185	....	300,000	.0365	....
70,000	.0075	....	5,000	....	.0060	310,000	.0392	....
80,000	.0080	....	200,000	.0192	....	319,000	.0450	....
90,000	.0090	....	220,000	.0215	....	320,000	.0550	....
100,000	.0098	....	230,000	.0225	....	321,200	.0600	broken

SPECIAL TABLE IV.—(*Concluded.*)

16-INCH MORTAR CUBE, MARKED *Ab*; BEDS PLASTERED.

Actual size: Bed = 16".05 × 16".08; Height = 16".08 (or 16".17 including plaster); Weight, 284½ pounds.

| LOAD. | INCH. | | LOAD. | INCH. | | LOAD. | INCH. | |
Pounds.	Compression.	Set.	Pounds.	Compression.	Set.	Pounds.	Compression.	Set.
5,000	....	....	5,000	....	.0025	240,000	.0230	....
10,000	.0005	....	100,000	.0090	....	250,000	.0245	....
20,000	.0015	....	120,000	.0105	....	260,000	.0262	....
30,000	.0027	....	140,000	.0120	....	270,000	.0288	....
40,000	.0040	....	160,000	.0138	....	280,000	.0310	....
50,000	.0050	....	180,000	.0155	....	290,000	.0340	....
60,000	.0060	....	200,000	.0175	....	300,000	.0390	....
70,000	.0070	....	5,000	....	.0052	310,000	.0445	....
80,000	.0075	....	200,000	.0182	....	320,000	.0520	broken
90,000	.0080	....	220,000	.0200	....	....	....	....
100,000	.0090	....	230,000	.0215	....	....	....	....

SPECIAL TABLE V.

SHOWING AMOUNT OF COMPRESSION AND SET OF CUBES OF CONCRETE MADE WITH NORTON'S CEMENT.

Composition : 1 vol. Cement Paste, 1½ vols. Sand, and 6 vols. Broken Stone.

8-INCH CONCRETE CUBE, MARKED *Aa*; BEDS PLASTERED.

Actual size: Bed = 8".03 × 8".07; Height = 8".06 (or 8".16 including plaster); Weight, 43½ pounds.

| LOAD. | INCH. | | LOAD. | INCH. | | LOAD. | INCH. | |
Pounds.	Compression.	Set.	Pounds.	Compression.	Set.	Pounds.	Compression.	Set.
1,000	....	....	30,000	.0070	....	60,000	.0175	... ...
5,000	.0025	....	35,000	.0080	... .	65,000	.0215	....
10,000	.0030	....	40,000	.0090	....	70,000	.0260	....
15,000	.0042	....	45,000	.0105	....	74,300	.0310	....
20,000	.0050	....	50,000	.0122	....	75,000	.0325	....
25,000	.0060	....	1,000	....	.0065	80,000	.0385	....
1,000	....	.0030	50,000	.0130	....	85,000	.0485	....
25,000	.0062	....	55,000	.0145	....	87,600	.0690	broken

SPECIAL TABLE V. —(*Continued.*)

8-Inch Concrete Cube, marked *Ab*; Beds Plastered.

Actual size: Bed = 8".05 × 8".04; Height=8".04 (or 8".07 including plaster); Weight, 43 pounds.

Load.	Inch.		Load.	Inch.		Load.	Inch.	
Pounds.	Compression.	Set.	Pounds.	Compression.	Set.	Pounds.	Compression.	Set.
1,000	....	....	35,000	.0140	....	70,000	.0310	....
5,000	.0040	....	40,000	.0150	....	75,000	.0350	....
10,000	.0070	....	45,000	.0170	... .	80,000	.0398	....
15,000	.0085	....	50,000	.0190	....	85,000	.0450	....
20,000	.0098	....	1,000	....	.0130	90,000	.0575	.. ....
25,000	.0110	....	50,000	.0200	....	95,000	.0710	....
1,000	....	.0078	55,000	.0220	....	97,900	.1000	broken
25,000	.0115	....	60,000	.0240	....	....	....	....
30,000	.0120	....	65,000	.0275	....	....	....	....

12-Inch Concrete Cube, marked *Aa*; Beds Plastered.

Actual size: Bed = 12".09 × 12".00; Height = 12".02 (or 12".12 including plaster); Weight, 148 pounds.

Load.	Inch.		Load.	Inch.		Load.	Inch.	
Pounds.	Compression.	Set.	Pounds.	Compression.	Set.	Pounds.	Compression.	Set.
5,000	....	....	80,000	.0080	....	160,000	.0265	....
10,000	.0010	....	90,000	.0095	....	170,000	.0310	....
20,000	.0020	....	100,000	.0110	....	180,000	.0362	....
30,000	.0030	....	5,000	....	.0052	184,000	cracks in	sight
40,000	.0040	....	100,000	.0120	....	190,000	.0415	....
50,000	.0050	....	110,000	.0130	....	200,000	.0510	....
5,000	....	.0022	120,000	.0150	....	210,000	.0645	....
50,000	.0050	....	130,000	.0170	....	215,400	.0870	....
60,000	.0060	....	140,000	.0198	....	218,100	broken	....
70,000	.0070	....	150,000	.0230	....	....	....	....

SPECIAL TABLE V.—(*Continued.*)

12-Inch Concrete Cube, marked *Ab*; Beds Plastered.

Actual size: Bed = 12″.00 × 12″.00; Height = 12″.05 (or 12″.15 including plaster); Weight, 148½ pounds.

Load.	Inch.		Load.	Inch.		Load.	Inch.	
Pounds.	Compression.	Set.	Pounds.	Compression.	Set.	Pounds.	Compression.	Set.
5,000	....	....	90,000	.0148	....	150,000	.0292	....
10,000	.0015	....	100,000	.0160	....	160,000	.0320	....
20,000	.0042	....	5,000	....	.0098	170,000	.0340	....
30,000	.0058	....	100,000	.0170	....	180,000	.0380	....
50,000	.0085	....	110,000	.0185	....	190,000	.0430	cracks in sight
5,000	....	.0050	120,000	.0200	....	200,000	.0500	
50,000	.0087	....	130,000	.0220	....	210,000	.0590	....
60,000	.0100	....	140,000	.0245	....	220,000	.0720	....
70,000	.0115	....	150,000	.0280	....	228,300	.1100	....
80,000	.0130	....	5,000	....	.0172	232,900	broken	....

16-Inch Concrete Cube, marked *Aa*, 134; Beds Plastered.

Actual size: Bed = 16″.10 × 16″.07; Height = 16″.05 (or 16″.20 including plaster); Weight, 353 pounds.

Load.	Inch.		Load.	Inch.		Load.	Inch.	
Pounds.	Compression.	Set.	Pounds.	Compression.	Set.	Pounds.	Compression.	Set.
5,000	....	....	120,000	.0090	....	270,000	.0315	snappi'g sounds
10,000	.0008	....	140,000	.0100	....	280,000	.0360	....
20,000	.0030	....	160,000	.0115	....	290,000	.0400	....
30,000	.0042	....	180,000	.0130	....	300,000	.0450	....
40,000	.0049	....	200,000	.0150	....	310,000	.0500	cracks in sight
50,000	.0054	....	5,000	....	.0072	320,000	.0600	....
60,000	.0060	....	200,000	.0160	....	330,000	.0710	....
70,000	.0065	....	210,000	.0170	....	340,000	.0805	....
80,000	.0070	....	220,000	.0182	....	350,000	.0900	....
90,000	.0075	....	230,000	.0202	....	360,000	.1090	....
100,000	.0080	....	240,000	.0222	....	370,000	.1450	....
5,000	....	.0044	250,000	.0250	....	379,200	.2030	broken
100,000	.0080	....	260,000	.0275	....	....	....	....

SPECIAL TABLE V.—(*Concluded.*)

16-Inch Concrete Cube, marked *Ab*, 135; Beds Plastered.

Actual size: Bed = 16″.04 x 16″.05; Height = 16″.10 (or 16″.27 including plaster); Weight, 352½ pounds.

Load.	Inch.		Load.	Inch.		Load.	Inch.	
Pounds.	Compression.	Set.	Pounds.	Compression.	Set.	Pounds.	Compression.	Set.
5,000	....	....	120,000	.0080	....	280,000	.0320	....
10,000	.0008	....	140,000	.0092	....	290,000	.0360	....
20,000	.0015	....	160,000	.0110	....	300,000	.0420	....
30,000	.0025	....	180,000	.0130	....	310,000	.0500	....
40,000	.0030	....	200,000	.0150	....	318,700	.0538	....
50,000	.0038	....	5,000	....	.0070	320,000	.0580	....
60,000	.0042	....	200,000	.0160	....	330,000	.0602	....
70,000	.0050	....	220,000	.0180	....	340,000	.0650	....
80,000	.0055	....	230,000	.0192	....	350,000	.0740	....
90,000	.0060	....	240,000	.0208	....	360,000	.0875	....
100,000	.0069	....	250,000	.0235	....	368,000	.1170	broken
5,000	....	.0030	260,000	.0260	....	....	....	....
100,000	.0070	....	270,000	.0290	....	....	....	....

SPECIAL TABLE VI.

SHOWING AMOUNT OF COMPRESSION AND SET OF CUBES OF MORTAR MADE WITH NORTON'S CEMENT.

Composition : 1 vol. Cement Paste, 3 vols. Sand.

8-Inch Mortar Cube, marked *Ba*; Beds Plastered.

Actual size: Bed = 7″.96 x 8″.04; Height = 8″.05 (or 8″.18 including plaster); Weight, 35 pounds.

Load.	Inch.		Load.	Inch.		Load.	Inch.	
Pounds.	Compression.	Set.	Pounds.	Compression.	Set.	Pounds.	Compression.	Set.
1,000	....	....	25,000	.0090	....	40,000	.0150	....
5,000	.0025	....	1,000	....	.0030	45,000	.0180	....
10,000	.0042	....	25,000	.0095	....	50,000	.0230	....
15,000	.0060	....	30,000	.0110	....	54,250	.0400	broken
20,000	.0075	....	35,000	.0130	....	....	....	....

APPENDIX.

SPECIAL TABLE VI.—*(Continued.)*

8-INCH MORTAR CUBE, MARKED *Bb*; BEDS PLASTERED.

Actual size: Bed = 8".05 × 8".02; Height = 8".00 (or 8".10 including plaster); Weight, 35 pounds.

LOAD.	INCH.		LOAD.	INCH.		LOAD.	INCH.	
Pounds.	Compression.	Set.	Pounds.	Compression.	Set.	Pounds.	Compression.	Set.
1,000	....	....	25,000	.0090	....	40,000	.0160	....
5,000	.0030	....	1,000	....	.0040	45,000	.0230	....
10,000	.0050	....	25,000	.0095	....	47,250	.0360	broken
15,000	.0060	....	30,000	.0110	....	....	....	....
20,000	.0075	....	35,000	.0130	....	....	....	....

12-INCH MORTAR CUBE, MARKED *Ba*; BEDS PLASTERED.

Actual size: Bed = 12".02 × 12".06; Height = 12".00 (or 12".08 including plaster); Weight, 116 pounds.

LOAD.	INCH.		LOAD.	INCH.		LOAD.	INCH.	
Pounds.	Compression.	Set.	Pounds.	Compression.	Set.	Pounds.	Compression.	Set.
5,000	....	....	50,000	.0069	....	80,000	.0135	....
10,000	.0012	....	5,000	....	.0029	90,000	.0180	....
20,000	.0030	....	50,000	.0072	....	98,500	.0410	broken
30,000	.0040	....	60,000	.0083	....	....	....	....
40,000	.0055	....	70,000	.0108	....	....	....	....

12-INCH MORTAR CUBE, MARKED *Bb*; BEDS PLASTERED.

Actual size: Bed = 12".07 × 12".11; Height = 12".11 (or 12".14 including plaster); Weight, 116½ pounds.

LOAD.	INCH.		LOAD.	INCH.		LOAD.	INCH.	
Pounds.	Compression.	Set.	Pounds.	Compression.	Set.	Pounds.	Compression.	Set.
5,000	....	....	50,000	.0070	....	80,000	.0152	....
10,000	.0010	....	5,000	....	.0028	90,000	.0210	....
20,000	.0025	....	50,000	.0075	....	100,000	.0320	....
30,000	.0040	....	60,000	.0090	....	101,600	.0410	broken
40,000	.0055	....	70,000	.0120	....	....	....	....

SPECIAL TABLE VI.—(*Concluded.*)

16-Inch Mortar Cube, marked *Ba*; Beds Plastered.

Actual size: Bed = 16".10 × 16".11; Height = 16".10 (or 16.24 including plaster); Weight, 277½ pounds.

Load.	Inch.		Load.	Inch.		Load.	Inch.	
Pounds.	Compression.	Set.	Pounds.	Compression.	Set.	Pounds.	Compression.	Set.
5,000	....	....	80,000	.0082	....	140,000	.0180	....
10,000	.0010	....	90,000	.0095	....	150,000	.0205	....
20,000	.0025	....	100,000	.0110	....	160,000	.0235	....
30,000	.0035	....	5,000	.. ....	.0042	170,000	.0272	....
40,000	.0042	....	100,000	.0112	....	180,000	.0320	....
50,000	.0052	....	110,000	.0125	....	190,000	.0420	....
60,000	.0065	....	120,000	.0140	....	194,200	.0560	broken
70,000	.0075	....	130,000	.0160	....	....	....	....

16-Inch Mortar Cube, marked *Bb*; Beds Plastered.

Actual size: Bed = 16".07 × 16".00; Height = 16".09 (or 16".25 including plaster); Weight, 277½ pounds.

Load.	Inch.		Load.	Inch.		Load.	Inch.	
Pounds.	Compression.	Set.	Pounds.	Compression.	Set.	Pounds.	Compression.	Set.
5,000	....	....	70,000	.0098	....	120,000	.0172	....
10,000	.0020	....	80,000	.0110	....	130,000	.0192	....
20,000	.0038	....	90,000	.0122	....	140,000	.0225	....
30,000	.0050	....	100,000	.0140	....	150,000	.0258	....
40,000	.0062	....	5,000	....	.0055	160,000	.0302	....
50,000	.0072	....	100,000	.0145	....	170,000	.0380	....
60,000	.0082	....	110,000	.0160	....	176,750	.0540	broken

SPECIAL TABLE VII.

SHOWING AMOUNT OF COMPRESSION AND SET OF CUBES OF CONCRETE MADE WITH NORTON'S CEMENT.

Composition : 1 vol. Cement, 3 vols. Sand, 6 vols. Broken Stone.

8-INCH CONCRETE CUBE, MARKED *Ba* ; BEDS PLASTERED.

Actual size: Bed = 8″.02 × 8″.00; Height = 8″.02 (or 8″.16 including plaster); Weight, 42 pounds.

LOAD.	INCH.		LOAD.	INCH.		LOAD.	INCH.	
Pounds.	Compression.	Set.	Pounds.	Compression.	Set.	Pounds.	Compression.	Set.
1,000	....	....	25,000	.0142	....	40,000	.0215	....
5,000	.0062	....	1,000	....	.0110	45,000	.0258	....
10,000	.0090	....	25,000	.0150	....	50,000	.0330	....
15,000	.0110	....	30,000	.0162	....	54,300	.0480	....
20,000	.0120	....	35,000	.0185	....	56,400	broken	....

8-INCH CONCRETE CUBE, MARKED *Bb* ; BEDS PLASTERED.

Actual size: Bed = 8″.00 × 8″.15; Height = 8″.05 (or 8″.24 including plaster); Weight, 42½ pounds.

LOAD.	INCH.		LOAD.	INCH.		LOAD.	INCH.	
Pounds.	Compression.	Set.	Pounds.	Compression.	Set.	Pounds.	Compression.	Set.
1,000	....	....	25,000	.0085	....	40,000	.0142	....
5,000	.0022	....	1,000	....	.0042	45,000	.0172	....
10,000	.0040	....	25,000	.0090	....	50,000	.0230	....
15,000	.0055	....	30,000	.0100	....	55,000	.0450	broken
20,000	.0070	....	35,000	.0120	....	....	....	....

12-INCH CONCRETE CUBE, MARKED *Ba* ; BEDS PLASTERED.

Actual size: Bed = 12″.01 × 12″.11; Height = 12″.03 (or 12″.17 including plaster); Weight, 140 pounds.

LOAD.	INCH.		LOAD.	INCH.		LOAD.	INCH.	
Pounds.	Compression.	Set.	Pounds.	Compression.	Set.	Pounds.	Compression.	Set.
5,000	....	....	50,000	.0065	....	80,000	.0125	....
10,000	.0007	....	5,000	....	.0030	90,000	.0180	....
20,000	.0022	....	50,000	.0070	....	100,000	.0290	....
30,000	.0040	....	60,000	.0080	....	110,000	.0575	....
40,000	.0050	....	70,000	.0104	....	112,650	.0760	broken

SPECIAL TABLE VII.—(*Continued.*)

12-INCH CONCRETE CUBE, MARKED *Bb*; BEDS PLASTERED.

Actual size: Bed = 12".06 × 12".05; Height = 12".05. (or 12".14 including plaster); Weight, 140 pounds.

Load.	Inch.		Load.	Inch.		Load.	Inch.	
Pounds.	Compression.	Set.	Pounds.	Compression.	Set.	Pounds.	Compression.	Set.
5,000	....	....	50,000	.0062	....	80,000	.0112	....
10,000	.0010	....	5,000	....	.0025	90,000	.0135	....
20,000	.0025	....	50,000	.0062	....	100,000	.0240	....
30,000	.0038	....	60,000	.0075	....	109,900	.0525	broken
40,000	.0049	....	70,000	.0090	....	....	....	....

16-INCH CONCRETE CUBE, MARKED *Ba*; BEDS PLASTERED.

Actual size: Bed = 16".14 × 16".03; Height = 16".12 (or 16".21 including plaster); Weight, 339 pounds.

Load.	Inch.		Load.	Inch.		Load.	Inch.	
Pounds.	Compression.	Set.	Pounds.	Compression.	Set.	Pounds.	Compression.	Set.
5,000	....	....	100,000	.0123	....	220,000	.0500	....
10,000	.0008	....	110,000	.0130	....			Compression after sustaining load 5 minutes; cracks in sight.
20,000	.0022	....	120,000	.0140	....			
30,000	.0040	....	130,000	.0150	....			
40,000	.0050	....	140,000	.0170	....	222,100	.0660	
50,000	.0062	....	150,000	.0180	....			
5,000	....	.0045	5,000	....	.0120			
50,000	.0065	....	150,000	.0190	....	....	....	
60,000	.0075	....	160,000	.0200	....	222,100	.0940	After 10 minutes.
70,000	.0084	....	170,000	.0215	....	222,100	.1450	After 12
80,000	.0095	....	180,000	.0242	....	minutes, when disintegration		
90,000	.0105	....	190,000	.0272	....	took place rapidly.		
100,000	.0115	....	200,000	.0360	....			
5,000	....	.0072	210,000	.0450	....			

SPECIAL TABLE VII.—(*Concluded.*)

16-INCH CONCRETE CUBE, MARKED *Bb*; BEDS PLASTERED.

Actual size: Bed = 16".12 × 16".10; Height = 16".14 (or 16".24 including plaster); Weight, 339½ pounds.

Load.	Inch.		Load.	Inch.		Load.	Inch.	
Pounds.	Compression.	Set.	Pounds.	Compression.	Set.	Pounds.	Compression.	Set.
5,000	....	....	90,000	.0075	....	150,000	.0162	....
10,000	.0010	....	100,000	.0080	....	160,000	.0180	....
20,000	.0020	....	5,000	....	.0035	170,000	.0210	....
30,000	.0030	....	100,000	.0085	....	180,000	.0260	....
40,000	.0040	....	110,000	.0092	....	190,000	.0320	....
50,000	.0045	....	120,000	.0105	....	200,000	.0400	....
5,000	....	.0020	130,000	.0120	....	210,000	.0540	cracks in sight.
56,000	.0047	....	140,000	.0132	....	215,000	.0820	broken
60,000	.0060	....	150,000	.0150	....	....	....	....
80,000	.0068	....	5,000	....	.0070	....	....	....

SPECIAL TABLE VIII.

SHOWING AMOUNT OF COMPRESSION AND SET OF CUBES OF MORTAR MADE WITH NATIONAL PORTLAND CEMENT.

Composition: 1 vol. Cement Paste, 3 vols. Sand.

8-INCH MORTAR CUBE, MARKED *Ca*; BEDS PLASTERED.

Actual size: Bed = 8".08 × 8".04; Height = 8".01 (or 8".13 including plaster); Weight, 35½ pounds.

Load.	Inch.		Load.	Inch.		Load.	Inch.	
Pounds.	Compression.	Set.	Pounds.	Compression.	Set.	Pounds.	Compression.	Set.
1,000	....	....	60,000	.0055	....	130,000	.0122	....
5,000	.0008	....	70,000	.0062	....	140,000	.0138	....
10,000	.0012	....	80,000	.0071	....	150,000	.0150	....
20,000	.0020	....	90,000	.0080	....	1,000	....	.0030
30,000	.0030	....	100,000	.0090	....	150,000	.0160	....
40,000	.0038	....	1,000	....	.0015	160,000	.0170	....
50,000	.0045	....	100,000	.0095	....	168,000	.0210	broken
1,000	....	.0005	110,000	.0102	....	....	....	....
50,000	.0045	....	120,000	.0112	....	....	....	....

SPECIAL TABLE VIII.—(*Continued.*)

8-Inch Mortar Cube, marked *Cb*; Beds Plastered.

Actual size: Bed = 8".01 × 7".96; Height = 8".13 (or 8".25 including plaster); Weight, 36 pounds.

Load. Pounds.	Inch. Compression.	Set.	Load. Pounds.	Inch. Compression.	Set.	Load. Pounds.	Inch. Compression.	Set.
1,000	....	....	50,000	.0100	....	110,000	.0165	....
5,000	.0030	....	60,000	.0110	....	120,000	.0180	....
10,000	.0045	....	70,000	.0120	....	130,000	.0198	....
20,000	.0065	....	80,000	.0130	....	140,000	.0220	....
30,000	.0075	... ...	90,000	.0140	....	150,000	.0250	....
40,000	.0088	....	100,000	.0152	....	1,000	....	.0090
50,000	.0100	....	1.000	....	.0045	150,000	.0310	broken
1,000	....	.0032	100,000	.0158	....	....	....	....

12-Inch Mortar Cube, marked *Ca*; Beds Plastered.

Actual size: Bed = 12".00 × 12".05; Height = 12".07 (or 12".15 including plaster); Weight, 125 pounds.

Load. Pounds.	Inch. Compression.	Set.	Load. Pounds.	Inch. Compression.	Set.	Load. Pounds.	Inch. Compression.	Set.
5,000	....	....	140,000	.0075	....	290,000	.0168	....
10,000	.0010	... .	160,000	.0082	....	300,000	.0178	....
20,000	.0015	....	180,000	.0090	....	5,000	....	.0048
40,000	.0023	....	200,000	.0102	....	300,000	.0180	....
60,000	.0040	....	5,000	....	.0031	310,000	.0190	....
80,000	.0048	....	200,000	.0102	....	320,000	.0200	....
100,000	.0058	....	220,000	.0115	....	330,000	.0210	....
5,000	....	.0029	240,000	.0125	....	340,000	.0222	....
100,000	.0062	....	260,000	.0140	... ...	350,000	.0242	....
120,000	.0070	....	280,000	.0156	....	357,400	.0272	broken

SPECIAL TABLE VIII.—(Continued.)

12-Inch Mortar Cube, marked C6; Beds Plastered.

Actual size: Bed = 12".02 × 12".00; Height = 12".10 (or 12".15 including plaster); Weight, 125½ pounds.

Load. Pounds.	Inch. Compression.	Set.	Load. Pounds.	Inch. Compression.	Set.	Load. Pounds.	Inch. Compression.	Set.
5,000	....	....	140,000	.0068	....	280,000	.0168	.. ....
10,000	.0002	....	160,000	.0078	....	290,000	.0180	....
20,000	.0009	....	180,000	.0090	....	300,000	.0190	....
40,000	.0020	....	200,000	.0102	....	5,000	....	.0050
60,000	.0029	... .	5,000	....	.0022	300,000	.0195	....
80,000	.0038	....	200,000	.0107	....	310,000	.0210	....
100,000	.0045	....	220,000	.0120	....	320,000	.0220	....
5,000	....	.0010	240,000	.0132	....	330,000	.0235	....
100,000	.0045	....	260,000	.0150	....	340,000	.0260	....
120,000	.0057	....	270,000	.0159	....	345,600	.0290	broken

16-Inch Mortar Cube, marked Ca; Beds Plastered.

Actual size: Bed = 16".12 × 16".12; Height = 16".22 (or 16".24 including plaster); Weight, 283 pounds.

Load. Pounds.	Inch. Compression.	Set.	Load. Pounds.	Inch. Compression.	Set.	Load. Pounds.	Inch. Compression.	Set.
5,000	....	....	240,000	.0064	....	500,000	.0160	....
10,000	.0002	....	260,000	.0070	....	5,000	....	.0070
20,000	.0002	....	280,000	.0075	....	500,000	.0165	....
40,000	.0008	....	300,000	.0080	....	520,000	.0172	....
60,000	.0012	....	5,000	....	.0038	540,000	.0181	....
80,000	.0020	....	300,000	.0085	....	560,000	.0188	....
100,000	.0026	....	320,000	.0090	....	580,000	.0202	....
5,000	....	.0015	340,000	.0096	....	600,000	.0215	....
100,000	.0028	....	360,000	.0102	....	5,000	....	.0095
120,000	.0031	....	380,000	.0110	....	600,000	.0230	....
140,000	.0037	....	400,000	.0118	. ...	610,000	.0235	....
160,000	.0042	....	5,000	....	.0052	620,000	.0242	....
180,000	.0048	....	400,000	.0120	....	630,000	.0250	....
200,000	.0050	....	420,000	.0125	....	640,000	.0260	....
5,000	....	.0025	440,000	.0132	....	650,000	.0272	broken
200,000	.0055	....	460,000	.0140	....	....	....	....
220,000	.0060	....	480,000	.0150	....	....	....	....

SPECIAL TABLE VIII.—(*Concluded.*)

16-INCH MORTAR CUBE, MARKED *Cb*; BEDS PLASTERED.

Actual size: Bed = 16".04 x 16".08; Height = 16".12 (or 16".20 including plaster); Weight, 283 pounds.

| LOAD. | INCH. | | LOAD. | INCH. | | LOAD. | INCH. | |
Pounds.	Compression.	Set.	Pounds.	Compression.	Set.	Pounds.	Compression.	Set.
5,000	....	....	300,000	.0102	....	520,000	.0208	....
10,000	.0002	....	5,000	....	.0032	540,000	.0220	....
20,000	.0005	....	300,000	.0102	....	560,000	.0230	....
40,000	.0013	....	340,000	.0118	....	580,000	.0242	....
80,000	.0030	....	380,000	.0132	....	600,000	.0255	....
100,000	.0035	....	400,000	.0142	....	5,000	....	.0080
5,000	....	.0015	5,000	....	.0042	600,000	.0265	....
100,000	.0035	....	400,000	.0148	....	610,000	.0275	....
140,000	.0050	....	420,000	.0152	....	620,000	.0284	....
180,000	.0062	....	440,000	.0162	....	630,000	.0292	....
200,000	.0070	....	460,000	.0170	....	640,000	.0304	....
5,000	....	.0022	480,000	.0180	....	650,000	.0330	....
200,000	.0070	....	500,000	.0190	....	654,500	.0350	broken
240,000	.0080	....	5,000	....	.0060	....	....	....
280,000	.0094	....	500,000	.0198	....	....	....	....

SPECIAL TABLE IX.

SHOWING AMOUNT OF COMPRESSION AND SET OF CUBES OF CONCRETE MADE WITH NATIONAL PORTLAND CEMENT.

Composition: 1 vol. Cement Paste, 3 vols. Sand, 6 vols. Broken Stone.

8-INCH CONCRETE CUBE, MARKED *Ca*; BEDS PLASTERED.

Actual size: Bed = 8".04 x 7".99; Height = 8".11 (or 8".24 including plaster); Weight, 43 pounds.

| LOAD. | INCH. | | LOAD. | INCH. | | LOAD. | INCH. | |
Pounds.	Compression.	Set.	Pounds.	Compression.	Set.	Pounds.	Compression.	Set.
1,000	....	....	60,000	.0085	....	130,000	.0160	....
5,000	.0040	....	70,000	.0095	....	140,000	.0175	....
10,000	.0045	....	80,000	.0102	....	150,000	.0195	....
20,000	.0057	....	90,000	.0112	....	160,000	.0220	....
30,000	.0065	....	100,000	.0120	....	170,000	.0255	....
40,000	.0070	....	1,000	....	.0055	180,000	.0300	....
50,000	.0080	....	100,000	.0125	....	190,000	.0365	....
1,000	....	.0042	110,000	.0132	....	196,500	.0480	broken
50,000	.0080	....	120,000	.0145	....	....	....	....

SPECIAL TABLE IX.—(Continued.)

8-INCH CONCRETE CUBE, MARKED Cb; BEDS PLASTERED.

Actual size: Bed = 8".05 × 8".03; Height = 8".18 (or 8".21 including plaster); Weight, 43 pounds.

LOAD. Pounds.	INCH. Compression.	Set.	LOAD. Pounds.	INCH. Compression.	Set.	LOAD. Pounds.	INCH. Compression.	Set.
1,000	....	....	60,000	.0072	....	130,000	.0200	....
5,000	.0010	....	70,000	.0082	....	140,000	.0225	....
10,000	.0020	....	80,000	.0095	....	150,000	.0250	....
20,000	.0032	....	90,000	.0110	....	160,000	.0275	....
30,000	.0042	....	100,000	.0125	....	170,000	.0310	....
40,000	.0052	....	1,000	....	.0048	180,000	.0350	....
50,000	.0062	....	100,000	.0132	....	190,000	.0415	....
1,000	....	.0020	110,000	.0150	....	193,500	.0480	broken
50,000	.0065	....	120,000	.0170	....	....	....	... ...

12-INCH CONCRETE CUBE, MARKED Ca; BEDS PLASTERED.

Actual size: Bed = 12".00 × 12".04; Height = 12".09 (or 12".19 including plaster); Weight, 143 pounds.

LOAD. Pounds.	INCH. Compression.	Set.	LOAD. Pounds.	INCH. Compression.	Set.	LOAD. Pounds.	INCH. Compression.	Set.
5,000	....	....	190,000	.0140	....	300,000	.0372	....
10,000	.0010	....	200,000	.0155	... . .	5,000	....	.0248
20,000	.0020	....	5,000	....	.0090	300,000	.0400	....
40,000	.0032	....	200,000	.0162	....	310,000	.0420	....
60,000	.0042	....	210,000	.0180	....	320,000	.0440	....
80,000	.0052	....	220,000	.0195	....	330,000	.0472	....
100,000	.0065	....	230,000	.0220	....	340,000	.0505	....
5,000	....	.0030	240,000	.0240	....	350,000	.0540	....
100,000	.0065	....	250,000	.0260	....	360,000	.0615	....
120,000	.0075	....	260,000	.0280	. ...	365,500	.0670	cracks in sight
140,000	.0085	....	270,000	.0300	....	367,000	.0720	broken
160,000	.0100	....	280,000	.0325	... ...	....	....	....
180,000	.0125	....	290,000	.0345	... ...	....	....	....

SPECIAL TABLE IX.—(*Continued.*)

12-INCH CONCRETE CUBE, MARKED $C6$; BEDS PLASTERED.

Actual size: Bed = 12".00 × 12".03; Height = 12".10 (or 12".18 including plaster); Weight, 143½ pounds.

LOAD.	INCH.		LOAD.	INCH.		LOAD.	INCH.	
Pounds.	Compression.	Set.	Pounds.	Compression.	Set.	Pounds.	Compression.	Set.
5,000	.....	.....	180,000	.0118	.....	320,000	.0230	.....
10,000	.0010	.....	200,000	.0130	.....	330,000	.0240	.....
20,000	.0023	.....	5,000	.....	.0060	340,000	.0260	.....
40,000	.0045	.....	200,000	.0130	.....	350,000	.0275	.....
60,000	.0058	.....	220,000	.0140	.....	360,000	.0292	.....
80,000	.0068	.....	240,000	.0150	.....	370,000	.0312	.....
100,000	.0078	.....	260,000	.0170	.....	380,000	.0345	.....
5,000	.....	.0040	280,000	.0180	.....	390,000	.0380	.....
100,000	.0080	.....	300,000	.0200	.....	400,000	.0400	.....
120,000	.0087	.....	5,000	.....	.0100	410,000	.0500	broken
140,000	.0098	.....	300,000	.0212	.....	.....	.....	.....
160,000	.0108	.....	310,000	.0222	.....	.....	.....	.....

16-INCH CONCRETE CUBE, MARKED $C8$; BEDS PLASTERED.

Actual size: Bed = 16".06 × 16".15; Height = 16".11 (or 16".19 including plaster); Weight, 345 pounds.

LOAD.	INCH.		LOAD.	INCH.		LOAD.	INCH.	
Pounds.	Compression.	Set.	Pounds.	Compression.	Set.	Pounds.	Compression.	Set.
5,000	.....	.....	340,000	.0120	.....	600,000	.0322	.....
10,000	.0002	.....	380,000	.0138	.....	610,000	.0340	.....
20,000	.0009	.....	400,000	.0148	.....	620,000	.0350	.....
40,000	.0018	.....	5,000	.....	.0060	640,000	.0365	.....
80,000	.0030	.....	400,000	.0152	.....	650,000	.0375	.....
100,000	.0035	.....	420,000	.0162	.....	660,000	.0390	.....
5,000	.....	.0020	440,000	.0170	.....	670,000	.0410	.....
100,000	.0037	.....	460,000	.0182	.....	680,000	.0436	.....
140,000	.0050	.....	480,000	.0198	.....	690,000	.0465	.....
180,000	.0062	.....	500,000	.0210	.....	700,000	.0502	.....
200,000	.0069	.....	5,000	.....	.0085	710,000	.0535	cracks in sight
5,000	.....	.0030	500,000	.0220	.....	720,000	.0560	.....
200,000	.0070	.....	520,000	.0231	.....	730,000	.0610	.....
240,000	.0080	.....	540,000	.0245	.....	738,000	.0710	.....
280,000	.0095	.....	560,000	.0260	.....	740,000	.0770	.....
300,000	.0102	.....	580,000	.0278	.....	747,000	.0820	broken
5,000	.....	.0042	600,000	.0300	.....	.....	.....	.....
300,000	.0105	.....	5,000	.....	.0132	.....	.....	.....

APPENDIX.

SPECIAL TABLE IX.—(*Concluded.*)

16-INCH CONCRETE CUBE, MARKED Cб, 175; BEDS PLASTERED.

Actual size: Bed = 16".17 × 16".08; Height = 16".16. (or 16".24 including plaster); Weight, 352 pounds.

LOAD.	INCH.		LOAD.	INCH.		LOAD.	INCH.	
Pounds.	Compression.	Set.	Pounds.	Compression.	Set.	Pounds.	Compression.	Set.
5,000	......	....	670,000	.0303	....	5,000	aft. 2 min.	.0410
10,000	.0005	....	680,000	.0315	....	5,000	" 4 "	.0405
20,000	.0012	....	690,000	.0330	....	5,000	" 6 "	.0405
40,000	.0020	....	700,000	.0348	....	100,000	.0500	....
80,000	.0030	....	710,000	.0358	....	200,000	.0570	....
100,000	.0039	....	720,000	.0365	....	300,000	.0610	....
5,000	......	.0020	730,000	.0375	....	400,000	.0650	....
100,000	.0040	....	740,000	.0390	....	500,000	.0685	....
140,000	.0050	....	750,000	.0405	....	600,000	.0710	....
180,000	.006c	....	760,000	.0430	....	700,000	.0740	....
200,000	.0065	....	770,000	.0448	....	800,000	.0810	....
5,000	......	.0030	780,000	.0465	....	......	......	aftersustaining this load for 2 minutes.
200,000	.0067	....	790,000	.0490	....			
240,000	.0075	....	795,000	.0510	....	800,000	.0850	for 4 min
280,000	.0088	....	800,000	.0530	....			
300,000	.0092	....	5,000	....	.0270	800,000	.0880	" 6 "
5,000	....	.0039	100,000	.0335	....	800,000	.0890	" 6 "
300,000	.0096	....	200,000	.0380	....	800,000	.0910	" 8 "
340,000	.0110	....	300,000	.0420	....	800,000	.0930	" 10 "
380,000	.0120	....	400,000	.0450	....	5,000	....	.0550
400,000	.0130	....	500,000	.0480	....	5,000	aft. 2 min.	.0535
5,000	....	.0050	600,000	.0510	....	5,000	" 4 "	.0532
400,000	.0135	....	700,000	.0540	cracks in sight	5,000	" 6 "	.0532
440,000	.0150	....	800,000	.0600	....	100,000	.0660	....
480,000	.0162	....	5,000	....	.0320	200,000	.0720	....
500,000	.0175	....	400,000	.0520	....	300,000	.0770	....
5,000	....	.0068	600,000	.0570	....	400,000	.0812	....
500,000	.0182	....	700,000	.0610	....	500,000	.0850	....
540,000	.0202	....	800,000	.0665	....	600,000	.0885	....
580,000	.0221	....	....	....	after sustaining this load for 2 min.	700,000	.0930	....
600,000	.0240	....	800,000	.0692		800,000	.1020	....
5,000	....	.0095				800,000	.1210	aftersustaining the maximum load for 2 minutes, when the piece rapidly failed and broke. Time from first application of maximum load to final failure, 1 hour 20 minutes.
600,000	.0258	....	800,000	.0720	for 4 min.			
610,000	.0268	....	800,000	.0730	for 6 min.			
620,000	.0272	....	800,000	.0740	for 8 min.			
630,000	.0280	....	800,000	.0752	for 10 m.			
650,000	.0290	....	5,000	....	.0415			
660,000	.0300	....						

SPECIAL TABLE X.

Showing Amount of Compression and Set of Short Solid Brick Piers.

Each pier was built of common, hard North River brick, in six courses, 1½ brick (or 12 inches) square in cross-section. The mortar consisted of 1 part Newark Co.'s Rosendale cement, and 2 parts sand. The mortar joints were about ⅜ inch thick. Each pier was furnished with base and cap of North River bluestone, and was made to represent ordinary brickwork.

First Brick Pier, marked I.; End Faces not Plastered.

Actual size: Section = $12''.00 \times 12''.00$; Length, brickwork, $16''.42$; including bluestone, $22''.42$.

Weight of brick only, 154 pounds; including bluestone, 238 pounds.

LOAD.	INCH.		LOAD.	INCH.		LOAD.	INCH.	
Pounds.	Compression.	Set.	Pounds.	Compression.	Set.	Pounds.	Compression.	Set.
5,000	.. ..	....	100,000	.0215	{ First snapp'g sound.	180,000	.0396	....
10,000	.0020	....	5,000	....	.0040	190,000	.0430	....
20,000	.0050	....	100,000	.0220	....	200,000	.0460	....
30,000	.0075	....	110,000	.0238	....	5,000	....	.0100
40,000	.0100	....	120,000	.0255	....	200,000	.0490	....
50,000	.0120	....	130,000	.0275	....	220,000	.0540	{ longitudinal
5,000	....	.0022	140,000	.0298	....	240,000	.0615	{ tudinal cracks in 2 courses
50,000	.0120	....	150,000	.0322	....	260,000	.0745	....
60,000	.0140	....	5,000	....	.0062	280,000	.0900	....
70,000	.0155	....	150,000	.0332	....	291,000	broken	....
80,000	.0175	....	160,000	.0352	....	....	....	....
90,000	.0192	....	170,000	.0370	....	....	....	....

SPECIAL TABLE X.—(*Continued.*)

SECOND BRICK PIER, MARKED II.; END FACES NOT PLASTERED.

Actual size: Section = 12″.00 × 11″.90; Length, brickwork, 16″.53; including bluestone, 22″.08.

Weight of brick only, 151 pounds; including bluestone, 233 pounds.

Load.	Inch.		Load.	Inch.		Load.	Inch.	
Pounds.	Compression.	Set.	Pounds.	Compression.	Set.	Pounds.	Compression.	Set.
5,000	.... ..	....	90,000	.0230	....	160,000	.0405	snappi'g sounds.
10,000	.0020	....	100,000	.0252	....	170,000	.0430	....
20,000	.0050	....	5,000	.. ..	.0050	180,000	.0460	....
30,000	.0080	....	100,000	.0260	....	190,000	.0498	....
40,000	.0108	....	110,000	.0278	....	200,000	.0530	....
50,000	.0132	....	120,000	.0302	....	5,000	....	.0132
5,000	....	.0030	130,000	.0330	....	200,000	.0552	....
50,000	.0138	....	140,000	.0350	....	220,000	.0600	cracks in 3 courses
60,000	.0150	....	150,000	.0375	....	240,000	.0720	....
70,000	.0180	....	5,000	....	.0081	260,000	.0940	broken
80,000	.0204	....	150,000	.0388	....	....	....	....

THIRD BRICK PIER, MARKED III.; END FACES NOT PLASTERED.

Actual size: Section = 12″.00 × 12″.00; Length, brickwork, 16″.32; including bluestone, 22″.58.

Weight of brickwork only, 154 pounds; including bluestone, 241 pounds.

Load.	Inch.		Load.	Inch.		Load.	Inch.	
Pounds.	Compression.	Set.	Pounds.	Compression.	Set.	Pounds.	Compression.	Set.
5,000	....	....	100,000	.0252	....	180,000	.0480	....
10,000	.0035	....	5,000	....	.0062	190,000	.0522	....
20,000	.0080	....	100,000	.0260	....	200,000	.0565	....
30,000	.0102	....	110,000	.0280	....	5,000	....	.0168
40,000	.0125	....	120,000	.0298	....	200,000	.0590	cracks in 3 courses
50,000	.0150	....	130,000	.0320	....	210,000	.0620	....
5,000	....	.0042	140,000	.0360	snapping sounds.	220,000	.0662	....
50,000	.0152	....	150,000	.0390	....	230,000	.0720	....
60,000	.0170	....	5,000	....	.0112	240,000	.0790	....
70,000	.0192	....	150,000	.0402	....	250,000	.0880	....
80,000	.0210	....	160,000	.0423	....	260,000	.1030	broken
90,000	.0230	.... .	170,000	.0450	....	....	....	....

SPECIAL TABLE X.—(*Continued.*)

FOURTH BRICK PIER, MARKED IV.; END FACES NOT PLASTERED.

Actual size: Section = 12".00 × 12".02; Length, brickwork, 16".25; including bluestone, 22".50. Weight of brickwork only, 153 pounds; including bluestone, 240 pounds.

LOAD.	INCH.		LOAD.	INCH.		LOAD.	INCH.	
Pounds.	Compression.	Set.	Pounds.	Compression.	Set.	Pounds.	Compression.	Set.
5,000	....	....	5,000	....	.0030	200,000	.0430	....
10,000	.0020	....	100,000	.0190	....	5,000	....	.0092
20,000	.0042	....	110,000	.0210	....	200,000	.0460	cracks in 2 courses
30,000	.0062	....	120,000	.0230	....			
40,000	.0080	....	130,000	.0250	....	210,000	.0485	....
50,000	.0100	....	140,000	.0270	....	220,000	.0512	....
5,000	....	.0020	150,000	.0295	....	230,000	.0550	....
50,000	.0100	....	5,000	....	.0050	240,000	.0600	....
60,000	.0115	....	150,000	.0310	....	250,000	.0650	....
70,000	.0130	....	160,000	.0330	....	260,000	.0745	....
80,000	.0150	....	170,000	.0352	....	270,000	.0870	....
90,000	.0170	....	180,000	.0370	....	280,000	.0990	....
100,000	.0190	....	190,000	.0400	snapping sounds	....	....	broken

FIFTH BRICK PIER, MARKED V.; END FACES NOT PLASTERED.

Actual size: Section = 12".00 × 12".00; Length, brickwork, 15".97; including bluestone, 23".22. Weight of brickwork only, 148 pounds; including bluestone, 251 pounds.

LOAD.	INCH.		LOAD.	INCH.		LOAD.	INCH.	
Pounds.	Compression.	Set.	Pounds.	Compression.	Set.	Pounds.	Compression.	Set.
5,000	....	....	100,000	.0320	....	180,000	.0570	....
10,000	.0040	....	5,000	....	.0110	190,000	.0610	... ...
20,000	.0095	....	100,000	.0330	....	200,000	.0660	3d course begins to flake off.
30,000	.0132	....	110,000	.0350	....			
40,000	.0160	....	120,000	.0375	....	5,000	....	.0228
50,000	.0192	....	130,000	.0410	....	200,000	.0688	....
5,000	....	.0072	140,000	.0435	....	210,000	.0740	....
50,000	.0195	....	150,000	.0470	....	220,000	.0785	*........
60,000	.0220	....	5,000	....	.0160	230,000	.0842	....
70,000	.0245	....	150,000	.0490	....	240,000	.0915	....
80,000	.0270	....	160,000	.0510	....	250,000	.1130	broken
90,000	.0295	....	170,000	.0540	....	*At 220,000 pds. general development of longitudinal cracks.		

SPECIAL TABLE X.—(*Concluded.*)

SIXTH BRICK PIER, MARKED VI.; END FACES NOT PLASTERED.

Actual size: Section = 12".00 × 11".75; Length, brickwork, 15".88; including bluestone, 21".98.
Weight of brickwork only, 147 pounds; including bluestone, 230 pounds.

LOAD.	INCH.		LOAD.	INCH.		LOAD.	INCH.	
Pounds.	Compression.	Set.	Pounds.	Compression.	Set.	Pounds.	Compression.	Set.
5,000	....	... .	100,000	.0225	....	180,000	.0470	....
10,000	.0030	....	5,000	....	.0040	190,000	.0510	....
20,000	.0060	....	100,000	.0235	....	200,000	.0550	....
30,000	.0080	....	110,000	.0252	....	5,000	....	.0148
40,000	.0102	....	120,000	.0275	....	200,000	.0590	....
50,000	.0120	....	130,000	.0302	....	210,000	.0632	....
5,000	....	.0022	140,000	.0330	....	220,000	.0700	....
50,000	.0120	....	150,000	.0360	....	230,000	.0760	....
60,000	.0140	....	5,000	....	.0080	240,000	.0870	....
70,000	.0160	....	150,000	.0372	....	250,000	.0990	cracks in 2d course
80,000	.0180	....	160,000	.0400	....			
90,000	.0205	....	170,000	.0432	....	251,000	.1090	broken

INDEX.

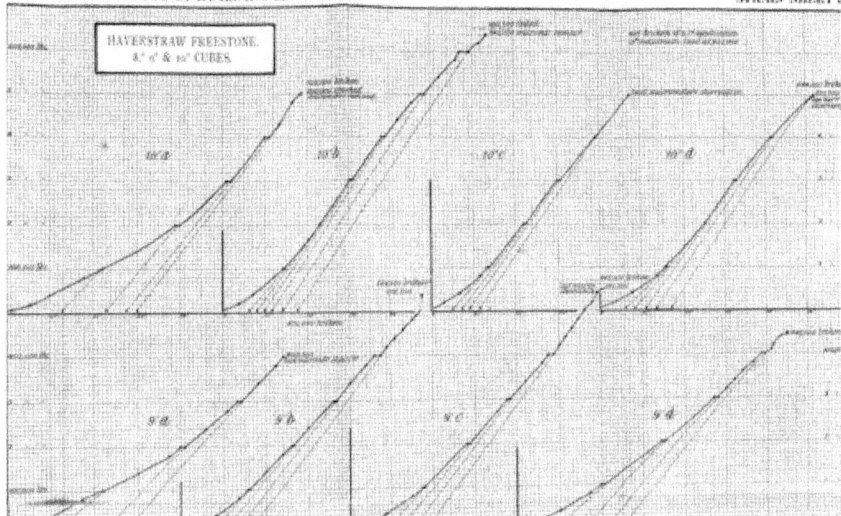

HAVERSTRAW FREESTONE.
8." 9" & 10" CUBES.

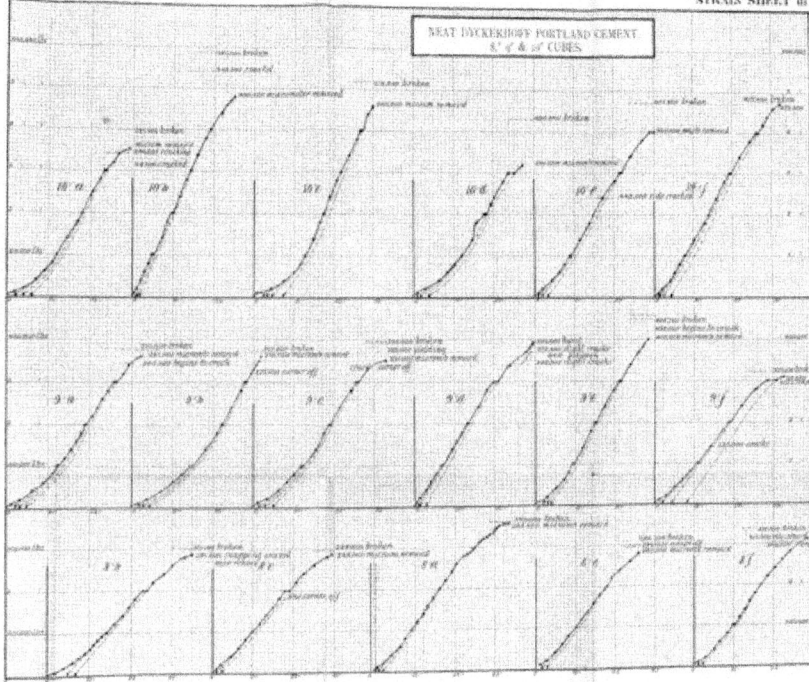

NEAT DYCKERHOFF PORTLAND CEMENT
8,' 4' & 20' CUBES

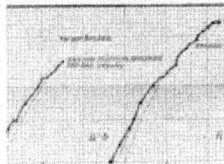

NEAT DYCKERHOFF PORTLAND CEMENT.
11″ & 12″ CUBES.

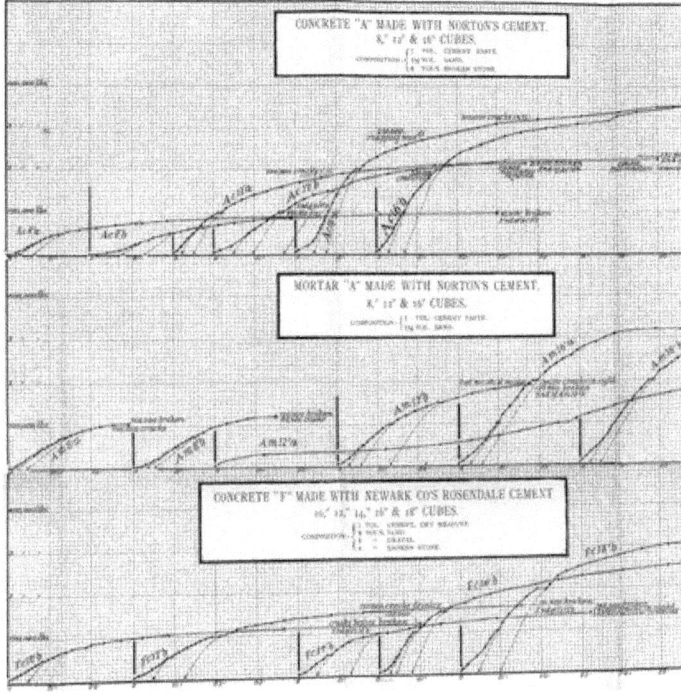

CONCRETE "A" MADE WITH NORTON'S CEMENT.
8," 12" & 18" CUBES.

MORTAR "A" MADE WITH NORTON'S CEMENT.
8," 12" & 16" CUBES.

CONCRETE "F" MADE WITH NEWARK CO'S ROSENDALE CEMENT.
10," 12," 14," 16" & 18" CUBES.

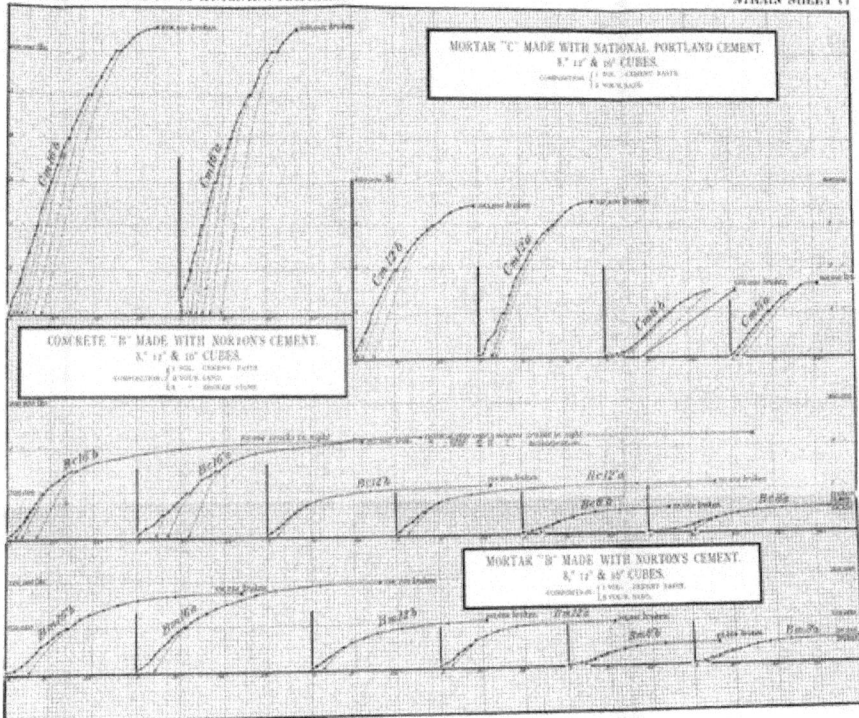

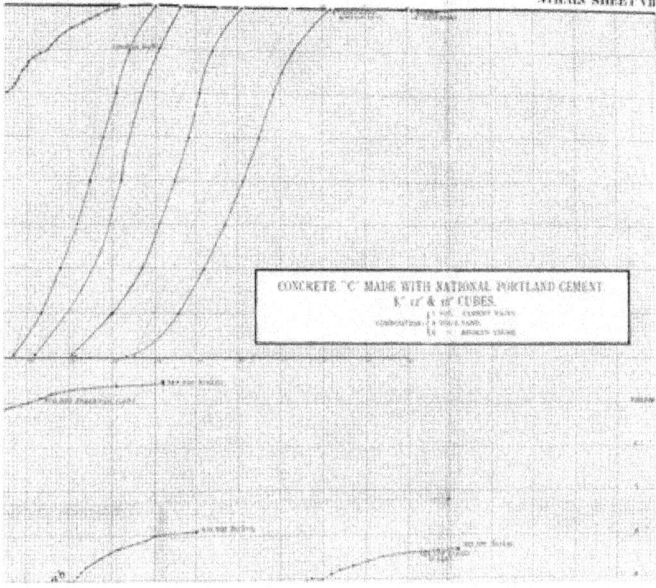

CONCRETE "C" MADE WITH NATIONAL PORTLAND CEMENT
12" & 18" CUBES.

RE IN CROSS-
COMMON HARD
NEWARK COM-
CH PIER WAS

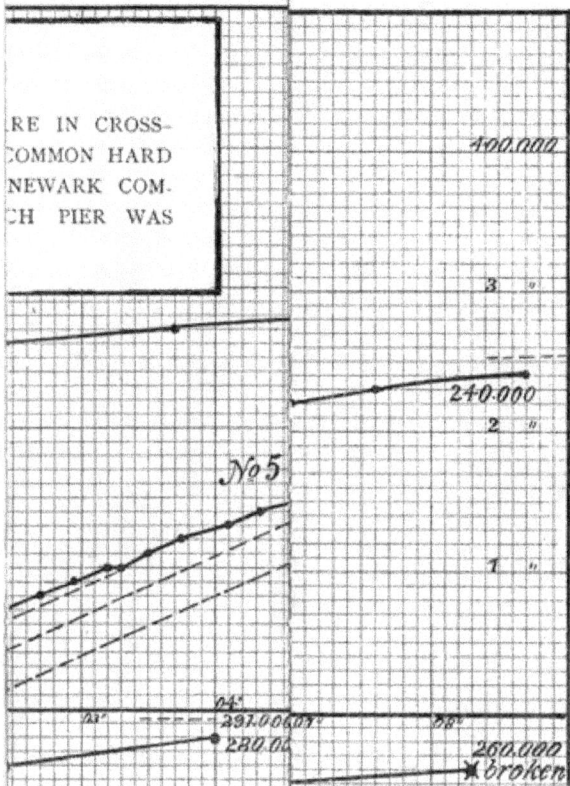

400.000

3

broken
251.000 at 1090
250.000 ι.0990
cracks ι.2ᵈ course

240.000

2

№5

1

04"

231.000

280.00

22"

28"

260.000
broken

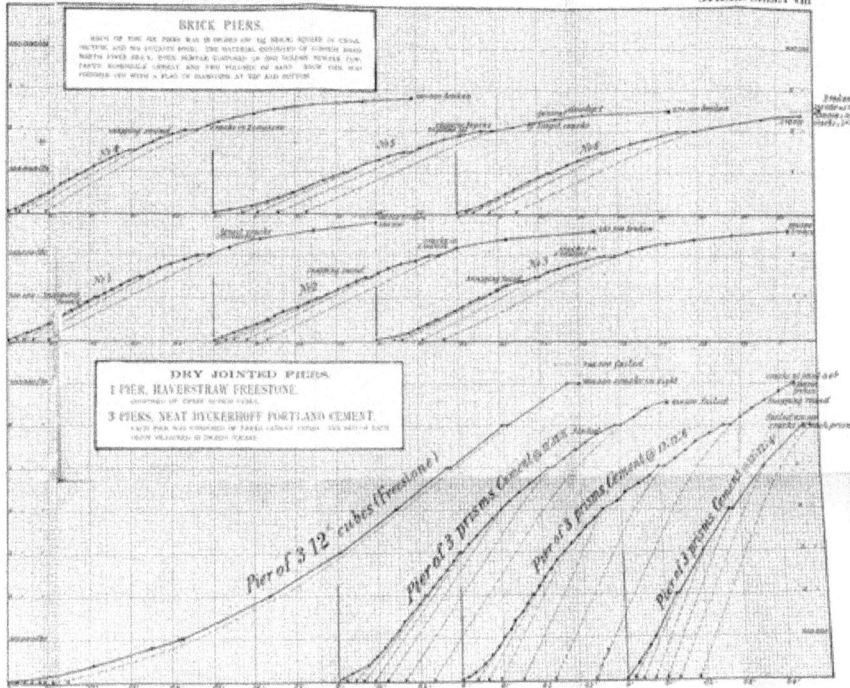

www.ingramcontent.com/pod-product-compliance
Lightning Source LLC
Chambersburg PA
CBHW021529210326
41599CB00012B/1431